Solving the Groundwater Challenges of the 21st Century

Selected papers on hydrogeology

22

Series Editor: Dr. Nick S. Robins
Editor-in-Chief IAH Book Series, British Geological Survey, Wallingford, UK

INTERNATIONAL ASSOCIATION OF HYDROGEOLOGISTS

Solving the Groundwater Challenges of the 21st Century

Editor

Ryan Vogwill

School of Earth and Environment, The University of Western Australia, Crawley, Australia

CRC Press
Taylor & Francis Group
Boca Raton London New York

CRC Press is an imprint of the
Taylor & Francis Group, an **informa** business

A BALKEMA BOOK

Published by: CRC Press/Balkema
               P.O. Box 11320, 2301 EH Leiden, The Netherlands
               e-mail: Pub.NL@taylorandfrancis.com
               www.crcpress.com – www.taylorandfrancis.com

First issued in paperback 2020

© 2016 Taylor & Francis Group, LLC
CRC Press/Balkema is an imprint of the Taylor & Francis Group, an informa business

No claim to original U.S. Government works

ISBN 13: 978-0-367-57486-4 (pbk)
ISBN 13: 978-1-138-02747-3 (hbk)

Visit the Taylor & Francis Web site at
http://www.taylorandfrancis.com

and the CRC Press Web site at
http://www.crcpress.com

Typeset by MPS Limited, Chennai, India

Library of Congress Cataloging-in-Publication Data

Table of contents

Solutions to the groundwater challenges of the 21st century

R. Vogwill

School of Earth and Environment, The University of Western Australia,
Perth, Australia

The Wagyl, according to local Western Australian indigenous Noongar culture, is a snakelike dreamtime creature responsible for the creation of waterways and landforms. This mythical being is strongly associated with rivers, lakes and is supposed still to reside deep beneath springs, effectively in the aquifer. As the Wagyl slithered over the land, his track shaped the sand dunes, his body scoured out the course of the rivers; where he occasionally stopped for a rest, he created bays and lakes. Outcrops of limestone are said to be his droppings. As he moved, his scales scraped off and become the forests and woodlands. As such all of these various sites are considered sacred by the local Noongar community.

The groundwater industry has huge challenges this century which mirror the issues in our global society in becoming sustainable. The book chapters are tailored to showcase solutions to the groundwater industry's biggest challenges. One of the key issues for all of society relates to changes in our future climate. Regardless of the causes, variability from the current patterns is likely. The prediction, possible mitigation and adaption to these impacts will be a crucial part of our success as a global society. Managed aquifer recharge is an important tool in managing climate change impacts as well as overabstraction. In his chapter Peter Dillon presents a review of hydrogeological issues in managed aquifer recharge and its role in solutions to global challenges.

In areas of rapid population growth in the developing world there are critical issues relating to sanitation and water supply. Stephen Foster explores these issues looking at solutions to these issues at a high level across all jurisdictions. Ali Bakari, a Nigerian PhD candidate, looks at this issue from a locals perspective, specifically in Nigeria, discussing how to overcome barriers to implementing sustainable groundwater management in a developing nation.

There is considerable risk of impacts to groundwater resources from extracting unconventional hydrocarbon deposits. This critical emerging issue has caused significant impacts to groundwater resources in some parts of the world already. Most jurisdictions are grappling with how best to manage the approval process for unconventional hydrocarbon projects and a critical component of managing this issue is getting regulation right. Louise Lennon and Ray Evans summarise the state of regulation of the unconventional hydrocarbon industry across Australia, identifying gaps and making recommendations critical to ensure groundwater resources in Australia are protected in this context. We believe this will be an invaluable summary for many other jurisdictions around the world to learn from and build on.

In many parts of the world there are legacies of various forms of contamination, impacting groundwater resources and the environment. Some of the most prevalent forms of groundwater contamination are agricultural chemicals and nutrients, given the large amount of agricultural land around the world. Niels Pedersen, Bo Vægter and their colleagues from Denmark present the results of their work over more than the last decade to identify areas of impact, areas at risk of impact, and how to manage this legacy of agricultural chemicals and nutrients going forward. This is an excellent example to many other jurisdictions with similar issues.

Last but definitely not least John Doherty's chapter that every modeller or user of modelling information must read about how we misuse models and how to use them better. This highly entertaining paper is a key part of the new modelling ethos that John has spearheaded over more than the last decade to make modelling more transparent and impartial. This is sorely needed. This ethos critically revolves around incorporating predictive uncertainty and probabilistic frameworks in groundwater modelling as opposed to 'cherry picking' a set of parameters which allow the model to draw the desired conclusion.

The chapters are selected from papers presented at the 40th IAH Congress – Solutions to the Groundwater Challenges of the 21st Century held at Perth.

Chapter 2

Managing aquifer recharge in integrated solutions to groundwater challenges

Peter Dillon

Co-Chair, IAH Commission on Managing Aquifer Recharge, Honorary Fellow, CSIRO Land and Water, Glen Osmond, SA, Australia

ABSTRACT

This chapter draws on recent scientific knowledge of aquifer processes in managed aquifer recharge to inform practical applications. Processes discussed include: clogging and its management in infiltration basins and injection wells, recovery efficiency of recharged water, and water quality changes in aquifers. This scientific information has been applied in managed aquifer recharge guidelines for protection of human health and the environment and supported a policy framework for groundwater entitlements. When these are adopted or adapted by water resources managers within local groundwater management plans this will: facilitate recovery of overexploited aquifers, reduce costs of water supply, improve groundwater quality, protect groundwater dependent ecosystems, support measures to rationalise water demand, supply new industry and agriculture, and/or progress long-term water banking as a buffer against future droughts. Networking and knowledge exchange would help governments and development banks to recognise circumstances where groundwater replenishment, is a credible and efficient supply or water security measure. Advances in groundwater replenishment including water banking will help maximise the benefits of aquifers to society and thereby assist water resources managers to address the water security challenges of the 21st century.

2.1 INTRODUCTION

Addressing groundwater overdraft is among the most pressing environmental and economic challenges globally and particularly in areas that have relied heavily on groundwater to support irrigated agriculture (Margat, 2008; Konikow, 2011). The enormous increase in productivity of irrigated land over dryland production in semi-arid areas has lifted communities out of hunger and poverty, and created investment in education, health, and economic infrastructure. Technologies of the 20th century such as drilling rigs, pumps, lightweight pipes, and distribution of electrical energy have enabled withdrawals of groundwater resources that have transformed the lives of hundreds of millions of rural families around the world. The human condition orients behaviour to increase utilisation of limiting resources so long as returns exceed costs. However, groundwater resources are finite. While resource utilisation has helped by producing wealth the consequent decline in groundwater levels, in some cases has impacted connected aquatic and riparian ecosystems, increased the costs of water access by farms and households, and ultimately led to reduced groundwater yields.

This has threatened livelihoods, impoverished health, impacted wellbeing and caused social disruption (Burke & Moench, 2000). Was the green revolution a once in an epoch windfall gain to one or two generations of people, and reversion to the former condition inevitable? This depends on our response to this challenge.

Few governments have demonstrated the will to curb current production to sustain a level of future production, which would be lower than that at present. Elected members of government who fail to acknowledge the problem nor support policies that preserve the productive value of groundwater, condemn their constituents to chronic or cataclysmic groundwater loss and consequent agricultural and social havoc. An informed community will not pass off declining groundwater levels and quality as an 'act of god' and understand that governments need to act decisively and fairly. Many farmers know that their children cannot make a living from a depleting aquifer so they continue to maximise irrigated production to enable their children to complete their education and take up city jobs that offer higher returns than farming. Their retirement income will be repatriated by their children supplemented by whatever can be gleaned from dryland farming or grazing. Farmers and governments do know that the windfall will end and that valued interventions are those that prolong the life of the water resource and improve the 'crop per drop' or commercial return per unit of water (CGIAR, 2015).

Investment in enhancing aquifer recharge is, therefore, a politically safe approach for governments. Recharge structures are a symbol of government support, and farmers are desirous of having these in close proximity to their wells (Maheshwari et al., 2014). However a supply-side strategy alone must be questioned in relation to effectiveness and equity of impact. Where groundwater withdrawal is two to 100 times the rate of natural recharge (Dillon et al., 2009a), we must ask frankly whether recharge enhancement really could make up this deficit. However in these extreme cases it can be a buffer that may prolong the viability of the resource long enough for some farmers to achieve their short-term objectives. More importantly, there is huge potential for recharge enhancement to be used as a lever to assist in water conservation (demand management), get much greater value out of government investment and increase farmer livelihood security. For example, formulation of a groundwater management plan, irrigation scheduling, mulching, soil testing, and even irrigation infrastructure improvements could be made a precondition for funding of recharge structures in an area as a package with agricultural extension services. Surprisingly this lever appears to be rarely used and could have a large aggregate impact on securing food production, reducing energy consumption, and stabilising rural communities.

If groundwater replenishment has so many benefits, why is it currently an underutilised tool in groundwater management? The barriers to adoption can include: lack of awareness of recent developments, inadequate approaches to groundwater protection and water entitlements: lack of knowledge of local aquifer systems, and lack of planning to give time to acquire such knowledge before water supply augmentation decisions are made. Managed aquifer recharge (MAR) is also an acid test of integrated water management, and often the fragmented state institutions, managing water resources and environmental impacts can complicate approvals. These impediments have been addressed in some jurisdictions and this message needs to be heard by governments and investors.

2.2 AQUIFER SUITABILITY FOR MAR

2.2.1 Overexploited aquifers

Aquifers that are already over-exploited are immediate potential targets for MAR. In these areas there is already a demonstration that an aquifer exists and that it has capacity for storing more water. The chief consideration is how to make MAR available as an inducement for the establishment and adherence to a groundwater management plan that: (a) engages with and accounts for all existing users of groundwater, (b) accounts for the sustainable rate of use of groundwater, with and without MAR, (c) establishes an equitable way of implementing water use efficiency and demand reduction measures that are verifiable and maximise the benefits to the community as a whole, (d) allows realistic estimates of achievable volume of MAR and a means of implementation where its cost is less than otherwise foregone production, and (e) restores the aquifer to hydrologic equilibrium over a finite time period to allow community adaptation towards a level of sustainable water use.

Table 2.1 illustrates the generalised suitability of aquifers for MAR, based on the properties of transmissivity, storage capacity, degree of containment, groundwater salinity and adherence to the relevant groundwater management plan. For over-exploited aquifers, generally the first four inherent properties, aquifer transmissivity, storage capacity, integrity of containment and freshness of ambient groundwater are not an impediment to the use of MAR. The most severe impediment is initial lack of adherence to a groundwater management plan, likely because no such plan exists that protects the aquifer from overdraft. While it is physically possible to implement MAR without a groundwater plan in place, the benefits are likely to be dissipated through inefficient water use, and may be localised, and so not evident to all groundwater users.

Establishing a groundwater management plan is essential to ensure that with sufficient investment in MAR (if suitable water sources are available) and with appropriate demand management, benefits will be maximised and equitably distributed. This means that, for example, the well owners closest to a recharge site will not extract all water newly available to them, in order that other groundwater users may achieve a benefit. In over-allocated systems, a pragmatic target should be to achieve seasonal storage rather than aim for water banking for the longer term. To achieve drought security would require a much more disciplined approach to groundwater use in areas where the starting position without MAR and demand management is unsustainable use.

Table 2.1 A generic guide to suitability of aquifers for managed aquifer recharge. Seasonal storage requires less onerous conditions than for inter-year storage, or water banking, intended to increase drought resilience and security of water supplies in a variable climate.

Aquifer property	Low	Medium	High
Transmissivity	Not suitable		
Storage capacity		Short term storage	Water banking
Degree of containment			
Groundwater freshness	Possible short term storage		
Groundwater plan adherence			

2.2.2   Aquifers initially in hydrologic equilibrium

For aquifers that are close to hydrologic equilibrium (not in a current state of overdraft) there are a number of factors affecting their suitability as targets of MAR. The most fundamental is the ability to recover recharged water. The proportion of recharged water that can be recovered for beneficial use is termed 'recovery efficiency'. Table 2.2 synthesises from a number of studies on recovery efficiency, including Pavelic *et al.* (2006); Ward *et al.* (2009), and Miotlinski *et al.* (2011) to illustrate impeding characteristics of aquifers and how to address these in practical ways.

Recovery efficiency in thermal energy storage differs from the recovery efficiency of conservative solutes in injectant (Miotlinski & Dillon, 2014) as the transport and

Table 2.2 Aquifer characteristics that may potentially impede recovery efficiency are identified along with preventive measures, operational practices, and monitoring that would be required in order to confidently manage these.

Aquifer characteristic	Preventative measures	Operational practices	Monitoring
High lateral groundwater gradient	Construct recovery wells downgradient	Store large volumes, if native groundwater of poor quality	Groundwater heads and quality monitored at observation piezometers
Leaky confining layer	Avoid recharge where confining layers are thin, or faulted	Maintain heads at MAR facility within a range that constrains leakage into and out of the aquifer	Groundwater heads and quality monitored at observation piezometers, including other side of and inside the confining layer
Fractured rock or karstic aquifer	Construct recovery wells downgradient	Store large volumes, if native groundwater of poor quality	Groundwater heads and quality monitored at observation piezometers, Tracer studies if required
Thin unconfined aquifer	Avoid site or horizontal recovery wells	Use low rates of recovery and recover to a surface storage if necessary	Groundwater heads and quality monitored at observation piezometers
Recharge close to aquifer discharge zone	Avoid site unless for saline intrusion barrier	Maintain head at lowest level that avoids saline intrusion (to avoid excessive loss)	Groundwater heads and quality monitored at observation piezometers and in discharge area
Brackish native groundwater	Avoid high salinity native groundwater in thick aquifers where density contrast affects flow in storage zone	Maximise storage volume, using multiple wells to create contiguous plume if necessary	Groundwater heads and quality monitored at observation piezometers. Monitor fluxes of water and salt recharged and recovered

mixing processes differ. However, conjunctive storage of water and thermal energy, is possible and warrants further evaluation, where this may contribute to the viability of MAR.

2.3 SOURCE WATER SUITABILITY FOR MAR

There is a plethora of experience on the use of a variety of sources of water for MAR. Natural waters have been recharged intentionally since before the 1960s (Johnson, 1988; Johnson & Pyne, 1994). Subsequently urban storm water and recycled water derived from sewage treatment plants have been successfully, intentionally used over three decades (Peters *et al.*, 1998; Dillon, 2002; Fritz *et al.*, 2005; Fox *et al.*, 2007; Herrman *et al.*, 2010). From these experiences two main issues emerge from the source of water used for recharge; clogging and safety of water quality for human health and the environment (including the aquifer).

2.3.1 Managing clogging

Managing soil or aquifer hydraulic conductivity is vitally important in managed aquifer recharge projects to ensure viability and longevity of operations. In the earlier years of MAR projects clogging of basins and wells was relatively common and in most cases was managed by operating procedures, such as scraping of basins and redevelopment of wells. Experience gained over the years has led to a greater emphasis on water treatment to reduce the frequency at which remedial measures are needed and improve cost-efficiency. Table 2.3 contains a simplified summary. For more information see Martin (ed.) (2013), Pyne (2005), Maliva & Missimer (2011), Pavelic *et al.* (2007, 2011), and Page *et al.* (2014). In injecting water into relatively inert aquifers (e.g. sands and sandstones) clogging preventive measures need to be more rigorous than for limestone aquifers where carbonate dissolution by organic matter in water helps to refresh the matrix surface and sustain hydraulic conductivity.

In a sense, clogging processes can act to help prevent groundwater contamination, but absence of clogging does not mean that the aquifer is adequately protected.

2.3.2 Protecting water quality

Many international research projects have been undertaken in the last decade or so by a range of research organisations. These have led to an improved understanding of the fate of constituents in water recharging aquifers and of the biogeochemistry of mixing within aquifers. Compilations including proceedings of International Symposia on MAR and research project summaries include: Johnson (ed.) (1988), Johnson & Pyne (eds.) (1994), Peters *et al.* (eds.) (1998), Dillon (ed.) (2002), Dillon & Toze (eds.) (2003), Fritz *et al.* (eds.) (2005), Fox *et al.* (2006, 2007), Nützmann *et al.* (eds.) (2006), Bixio & Wintgens (eds.) (2006), Jiménez & Asano (eds.) (2008), Bouwer *et al.* (2008), Van den Hoven & Kazner (eds.) (2009), Vanderzalm *et al.* (2009), Herrmann *et al.* (eds.) (2010), Ray & Shamrukh (eds.) (2011), Kazner, *et al.* (eds.) (2012), Sheng & Zhao (eds.) (2015), and Zhao & Wang (eds.) (2015).

By 2009 there was sufficient information to synthesise this knowledge into the first science-based MAR guidelines to manage health and environmental risks (NRMMC,

Table 2.3 Various types of clogging and their management in managed aquifer recharge.

Type of Clogging	Preventative measures	Remediation	Operational practices	Monitoring
Physical blocking by particulates	Treatments to remove particulates from source water	Purging wells, scraping basins	Source water controls, selective offtake	Specific capacity of injection well, infiltration rate of basin, turbidity of source water, frequency of remediation
Mobilisation of aquifer fine materials	Treatments to reduce sodium adsorption ratio (SAR) in source water	Purging wells, injecting divalent ions, scraping basins	Occasional dosing with divalent ions or specific polymers. Only gradual changes in injection rates	Specific capacity of injection well (continuously), infiltration rate of basin
Air entrainment, (gas binding)	Prevent air entry into injection wells	Purging wells, install air relief wells beneath basins	Air release valves, flow control valves, avoid cascading	Specific capacity of injection well, infiltration rate of basin (continuously), dissolved oxygen in injectant
Geochemical precipitation	Control precipitation potential in source water, e.g. through redox control, nutrient removal, or pH control	Purging wells, acid injection, scraping basins	Prevent unnecessary air entry for injection wells	Specific capacity of injection well, infiltration rate of basin, pH, Eh, Fe, Mn *etc.* of injectant and recovered water
Biofilm accumulation – organisms and polysaccharides	Treatments to reduce nutrients, e.g. labile organic carbon and prevent excess coagulant	Purging wells, scraping basins	Disinfection in wells in fine grained media, drying times in basins	Specific capacity of injection well, infiltration rate of basin
Combinations of the above	Relevant controls above	Purging wells, scraping basins	Relevant measures above	All relevant measures above

EPHC, NHMRC 2009). These Australian MAR Guidelines provide a framework to assess contaminants in source water and to evaluate the ability of the aquifer to remove them or detain them and to produce other contaminants through aquifer geochemical processes. This allows identification of any necessary additional water treatment processes to sustainably meet the requirements for intended uses of recovered water and for groundwater within the aquifer beyond an attenuation zone. It is expected that these pioneering guidelines will be periodically updated, based on on-going research results.

An aquifer's attributes, such as temperature, oxygen-reduction status, presence of reactive minerals, and fractures or macro-porosity can affect treatment capacity for any given contaminant or pathogen. Hence, site specific information is needed to tailor a treatment process that gives the required level of protection. Table 2.4 gives an overview of treatment via soil or aquifer passage during MAR, alongside other elements of a treatment train for water. It is necessary to know the end-use water quality requirements, source water quality, ambient groundwater quality, and aquifer characteristics to incorporate MAR effectively and safely, particularly for source waters of impaired quality. Over-treating water, has cost, energy and greenhouse gas penalties, and can also increase dissolution of aquifer minerals or starve microorganisms that are capable of degrading contaminants within the aquifer. A whole system perspective is needed and as research progresses, more definitive treatment trains can be developed for particular types of source waters and aquifer systems.

Generally a combination of processes is needed and the costs of these can be reduced by taking accurate account of treatment occurring through soil and/or aquifer passage. Currently, validation techniques need to be applied at each site at which reliance is made on the aquifer for effective treatment. Validation methods need to be applied broadly to build confidence in predicting soil or aquifer treatment. Potentially in future, credit could be assigned based on the growing knowledge base on removal

Table 2.4 Managing aquifer recharge for health and environmental risks requires an understanding of processes and rates of contaminant attenuation and generation in the soil and/or aquifer. This allows an optimal treatment train to be determined that achieves water quality objectives.

Water quality parameter	Removal effectiveness by water treatment process						
	Soil or aquifer passage	Settling pond or wetland	Rapid sand filtration	Coagulation and filtration	Granular activated carbon filtration	Micro-filtration	Reverse osmosis
pathogenic organisms	ineffective to effective	ineffective to effective	ineffective to effective	ineffective to effective	ineffective to effective	ineffective to effective	effective
inorganic chemicals	ineffective to effective	ineffective to effective	ineffective to effective	ineffective to effective	ineffective to effective	ineffective to effective	generally effective
salinity and sodicity	ineffective	ineffective	ineffective	ineffective	ineffective	ineffective	effective
nutrients (N, P, OC)	ineffective to effective	ineffective to effective	ineffective to effective	ineffective to effective	generally effective	ineffective to effective	effective
organic chemicals	ineffective to effective	ineffective to effective	ineffective	ineffective to effective	generally effective	ineffective	generally effective
turbidity and particulates	ineffective to effective	ineffective to effective	ineffective to effective	generally effective	ineffective	ineffective	ineffective
radionuclides	ineffective	ineffective to effective	ineffective	ineffective to effective	ineffective	ineffective to effective	effective

processes, rates of removal, and daughter products under the range of conditions normally encountered. These in general are simpler for natural waters used for recharge when the sole use of groundwater is for irrigation. Where groundwater is also used as a drinking water resource, protection needs to be afforded to drinking water wells with a pragmatic programme of assessment. A guideline for MAR (Dillon *et al.*, 2014) was developed for such circumstances in rural India. It relies on a sanitary survey and water safety plan, rather than sampling and detailed analysis of water quality, which is currently not logistically or financially feasible in most rural areas.

2.4 WATER BANKING – THE HIGHEST VALUE SOLUTION

In many areas climate change suggests that natural recharge may decline and demand for water will grow (e.g. in southern Australia, Barron *et al.*, 2011). This means that some systems, which until now have been in hydrologic equilibrium, may become over-allocated unless there is adaptive management in place. In order to sustain supplies and mitigate impacts, either groundwater extraction needs to be reduced or recharge needs to be augmented.

The value of water, reflected in its price, is higher in urban areas than in rural areas. Urban areas also have the advantage of additional sources of water that are sometimes unused, such as storm water and recycled water from sewage effluent, in either separate or combined systems. Hence, in urban areas MAR is generally more cost effective than suffering the problems of chronic water supply shortage. Due to the high value of urban land, infiltration basins are rarely used in cities and recharge systems there will typically use injection wells, such as aquifer storage and recovery (ASR) wells, where treated water is injected into and recovered from the same well. In rural areas the value of water for irrigation constrains expenditure on MAR and inexpensive infiltration basins can be cost-effective for augmenting groundwater by diverting occasional excess stream flows. There are a host of other methods of MAR available making use of local topography, geomorphology, and hydrological conditions (e.g. Bouwer, 2002; Tuinhof & Heederick, 2003; Dillon, 2005; Parsons *et al.*, 2012).

These same methods may be applied even in currently unstressed aquifers to bank water for the future to provide a buffer for drought (Steenbergen & Tuinhof, 2010; Dillon *et al.*, 2012; Tuinhof *et al.*, 2013; Ross, 2014). Aquifer storages have tradi-tionally played a sustaining role for water supplies (Foster & MacDonald, 2014), not being subject to the temporal vagaries of surface water availability. There is very high value associated with water supply security (Dillon, 2015), with examples given by Megdal & Dillon (eds.) (2015) and Arizona Water Banking Authority (2015).

Figure 2.1 is a diagrammatic representation of the effect of water banking and seasonal water storage policies for the same time series of surface water availability with recharge whenever this exceeds a threshold. In the water banking scenario the water accumulates as a reserve within the aquifer and is not drawn on until drought occurs when its recovery averts a water supply failure, with very substantial economic and social benefits. In the second scenario of seasonal storage and recovery, the reserve is drawn down each dry season to supplement supplies but the reserve is then inadequate to meet drought supplies. This illustrates that the value of the aquifer and the recharge system can be greatly enhanced if this sustaining supply role for the aquifer is bolstered

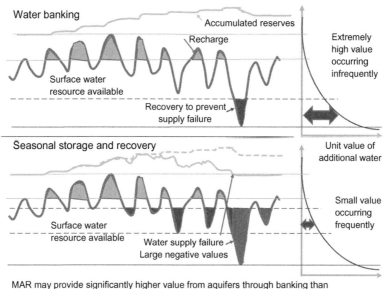

MAR may provide significantly higher value from aquifers through banking than through short term storage, but planning and proving investigations are needed.

Figure 2.1 Comparison of water banking and seasonal storage.

by water banking. As revealed in Tables 2.1 and 2.2, not all aquifers capable of seasonal storage are suitable for water banking. However, groundwater resources managers can use a portfolio approach, using each aquifer for the purpose to which it is best suited that yields the highest value.

2.5 SUBSURFACE STORAGE VERSUS SURFACE STORAGE

The advantages of water storage below ground over new dams are (Dillon *et al.*, 2009b):

– no evaporation, algae or mosquitoes,
– no prime productive valley floor land lost, nor population displaced,
– they are within or closer to places of high water demand than new dam sites,
– they have low capital costs, with smaller viable storages than dams,
– they are scale-able, allowing staged implementation, and,
– there are many more sites with suitable aquifers than suitable dam sites.

Dams and MAR are not mutually exclusive. Conjunctive operation can be highly efficient, where dams provide detention to allow time for recharge, which minimises evaporative losses over longer term storage periods.

The advantages of aquifers over dams increase in areas with high evaporation rates and low relief and in arid and semi-arid areas where storage capacity needs to be many times the mean annual demand in order to provide secure supplies. At one semi-arid site in Australia mean annual supply is less than 4% of the mean annual evaporation from the dam (SKM, 2010; Lawrie *et al.*, 2012). Higher recovery efficiencies have been observed for freshwater stored in even saline aquifers (Gerges *et al.*, 2002). As climate warms, evaporation will increase at an estimated 4% for every degree Celsius increase. It is prudent for hydrogeological evaluations to take place before making decisions on future water storages. Jurisdictions that ensure all drillers' logs, pumping test results, groundwater level measurements, and groundwater quality analyses are captured into a publically accessible data base will have an advantage when determining future water storage and supply options.

Lessons from dam construction over the last century need to be observed in advancing MAR. Firstly, the impact of harvesting water on flows downstream with attendant consequences for irrigation communities and aquatic ecosystems must be considered, regardless whether the storage is above or below ground. The solution to this is the formation of a catchment water allocation plan that safeguards all entitlements to water, including environmental entitlements. This has been documented in a policy framework for MAR that includes the incremental process of formulating such entitlements (Ward & Dillon, 2011).

Secondly, whereas in a dam there is undisputed entitlement to the stored water under an agreed plan for water sharing, this is less conspicuous in water stored in aquifers. Hence, there needs to be a water accounting procedure in place to know how much water has been recharged so the size of the pool of entitlements is defined. Defensible accounting can then be made for recoverability. In aquifers with finite storage capacity that capacity needs to be allocated among MAR operators and it is considered that having one MAR operator for each sector or for the whole of an aquifer is desirable in order to minimise dispute and maximise productive capacity of the system. Entitlements to recover water need to be transferable to facilitate co-investment by collectives of water users in MAR projects where they are most efficient, in proximity to sources of recharge water and to areas of water demand. These elements were included in the entitlement framework of Ward & Dillon (2011) and adopted in Western Australia (WA Department of Water, 2010).

2.6 CONCLUSIONS AND RECOMMENDATIONS

The scientific basis for MAR has been growing steadily over recent decades and recoverability of water, clogging, and water quality changes in the subsurface are now qualitatively predictable, although more research is needed to develop reliable quantitative tools to assist management of these issues. Techniques such as flood harvesting and water banking will provide challenges and benefits that warrant improved knowledge. Research is needed on the fate of bacteria, viruses, and parasites, such as *Cryptosporidium* in aquifers to enable recognition for the degree of their removal, accounting for inactivation, attachment, detachment, and infectivity. Field validation methods need to be developed. For developing countries using natural source waters, sanitation surveys and World Health Organisation compliant water safety plans will be useful in areas where water quality data are unavailable. Continued documentation of MAR, including post-audits of implementation experiences and economics are needed to give it the status it deserves on the water supply agenda, including irrigation, rural villages and peri-urban areas of growing cities. Improved forecasting of recharge credit depreciation for water banking operations and accounting for water quality are needed. Improved well design for saline intrusion barriers and to improve recovery in saline aquifers would also be useful.

Policy development and implementation is needed for MAR to achieve its potential role in integrated solutions to groundwater challenges. Of fundamental importance, MAR should be used as an inducement for the formation of water sharing plans for aquifers and implementation of demand management. Rights to additional supplies need to be equitably distributed, accounting for social values and spatially non-uniform aquifers, water supply sources, and demands within a community-supported groundwater management plan. Catchment water sharing plans are needed so that downstream users and ecosystems are not adversely impacted by MAR operations. Where aquifers are suitable, consideration should be given to advancing MAR from seasonal storage to water banking in a portfolio that sustains the resource and maximises the value of the aquifer and MAR operations.

Institutional reform and funding mechanisms are needed that encourage water banks, groundwater user associations, and water trading arrangements. In places where MAR is already practiced, better integration of approval processes for MAR are warranted, for example through formation of 'a one stop shop', a collective task force representing government agencies with conjoint responsibilities. Pilot projects are needed.

MAR is a field of scientific, technical, and policy endeavour that demands a greater knowledge of aquifers to gain a greater benefit from them. Those working on MAR are creating solutions for many of the challenges faced by humanity in the 21st century and beyond. This is just the beginning of the process of optimising the value of this suite of tools.

REFERENCES

Arizona Water Banking Authority (2015) http://www.azwaterbank.gov/

Barron O.V., Crosbie R.S., Charles S.P., Dawes W.R., Ali R., Evans W.R., Cresswell R., Pollock D., Hodgson G., Currie D., Mpelasoka F., Picket R., Aryal S., Donn M., Wurcher B. (2011). Climate change impact on groundwater resources in Australia. *Waterlines Report Series* No 67, Dec. 2011, National Water Commision. http://archive.nwc.gov.au/library/waterlines/67

Bixio D., Wintgens T. (eds.). (2006) Water Reuse System Management Manual: AQUAREC. Office for Official Publications of the European Communities.

Bouwer H. (2002) Artificial recharge of groundwater: hydrogeology and engineering. *Hydrogeology Journal* 10(1), 121–142.

Bouwer H., Pyne R., Brown A. (2008) *Design, operation and maintenance for sustainable underground storage facilities. American Water Works Association Research Foundation Report.* Denver, USA, 235p.

Burke J.J., Moench M. (2000) Groundwater and Society: Resources, Tensions and Opportunities. Themes in Groundwater Management for the Twenty-first Century. United Nations Department of Economic and Social Affairs and the Institute for Social and Environmental Transition, New York, USA.

CGIAR Research Program on Water, Land and Ecosystems (WLE) (2015) Groundwater and ecosystem services: a framework for managing smallholder groundwater-dependent agrarian socio-ecologies – applying an ecosystem services and resilience approach. Colombo, Sri Lanka: International Water Management Institute (IWMI). 25p. doi: 10.5337/2015.208

Dillon P.J. (ed.). (2002) *Management of Aquifer Recharge for Sustainability*. A.A. Balkema.

Dillon P., Toze S. (eds.). (2003) Water Quality Improvements During Aquifer Storage and Recovery. AWWARF Project 2618, Final Report.

Dillon P.J. (2005) Future management of aquifer recharge. Hydrogeology Journal, 13(1), 313–316.

Dillon P., Fernandez E.E., Tuinhof, A. (2012) Management of aquifer recharge and discharge processes and aquifer storage equilibrium. IAH contribution to GEF-FAO Groundwater Governance Thematic Paper 4, 49p. www.groundwatergovernance.org/resources/thematic-papers/en/

Dillon P., Gale I., Contreras S., Pavelic P., Evans R., Ward J. (2009a) *Managing aquifer recharge and discharge to sustain irrigation livelihoods under water scarcity and climate change*. IAHS Publication 330, 1–12.

Dillon P., Pavelic P., Page D., Beringen H., Ward J. (2009b) Managed Aquifer Recharge: An Introduction, Waterlines Report Series No 13, Feb 2009, National Water Commission. http://archive.nwc.gov.au/library/waterlines/13

Dillon P., Vanderzalm J., Sidhu J., Page D., Chadha D. (2014) A Water Quality Guide to Managed Aquifer Recharge in India. CSIRO Land and Water and UNESCO Report of AusAID PSLP Project ROU 14476. https://publications.csiro.au/rpr/pub?pid=csiro:EP149116

Dillon P.J. (2015) Australian progress in managed aquifer recharge and the water banking frontier. *Australian Water Association Journal Water* 42(6), 53–57.

Foster S., MacDonald A. (2014) The 'water security' dialogue: why it needs to be better informed about groundwater. *Hydrogeology Journal* 22, 1489–1492.

Fox P. *et al.* (2006) Advances in soil aquifer treatment: research for sustainable water reuse. American Water Works Association Research Foundation and Water Environment Research Foundation.

Fox P. (ed.). (2007) Management of Aquifer Recharge for Sustainability. Proceedings of ISMAR6. Amazon Press. www.recharge.iah.org/recharge/downloads/AquiferRecharge_ISMAR6.pdf

Fritz B. *et al.* (eds.) (2005) Recharge systems for protecting and enhancing groundwater resources. Proceedings of ISMAR5.IHP-VI, Series on Groundwater No. 13, UNESCO. www.unesdoc.unesco.org/images/0014/001492/149210E.pdf

Gerges N.Z., Howles S.R., Dennis K., Dillon P.J., Barry, K.E. (2002) Town water supply purification using aquifer storage and recovery in a saline aquifer. In *Management of Aquifer Recharge for Sustainability*, (ed. P.J. Dillon), A.A.Balkema. 459–464.

Herrman R. *et al.* (eds.). (2010) Achieving groundwater supply sustainability & reliability through managed aquifer recharge. Proceedings of ISMAR7. http://www.dina-mar.es/pdf/ismar7-proceedingsbook.pdf

Jiménez B., Asano T. (eds.) (2008) *Water Reuse: An International Survey of Current Practice, Issues and Needs*. IWA Publishing.

Johnson A.I. (ed.). (1988) Artificial Recharge of Groundwater. American Society of Civil Engineers, New York. http://cedb.asce.org/cgi/WWWdisplayrbn.cgi?RBN9780872627130

Johnson A.I., Pyne, R.D.G. (eds.). (1994) *Artificial Recharge of Groundwater II*. American Society of Civil Engineers, New York.

Kazner C., Wintgens T., Dillon P. (eds.). (2012) *Water Reclamation Technologies for Safe Managed Aquifer Recharge*. 429p. IWA Publishing, London, UK.

Konikow L.F. (2011) Contribution of global groundwater depletion since 1900 to sea-level rise. *Geophysical Research Letters* 38(L17401), 1–5.

Lawrie K.C. *et al.* (2012) Assessment of Conjunctive Water Supply Options to Enhance the Drought Security of Broken Hill, Regional Communities and Industries. Summary Report. Record 2012/15 Geocat # 73823, Report 5 of 5 Final Report July 2012.

Maheshwari B. M. *et al.* (2014) The role of transdisciplinary approach and community participation in village scale groundwater management: Insights from Gujarat and Rajasthan, India. *International Open Access Journal Water* 6(6), 3386–3408. http://www.mdpi.com/journal/water/special_issues/MAR

Maliva R.G., Missimer T.M. (2011) Aquifer storage and recovery and managed aquifer recharge using wells: planning, hydrogeology, design and operation. Schlumberger Water Services. Methods in Water Resources Evaluation Series No. 2.

Margat J. (2008) Les eaux souterraines dans le monde. BRGM Editions and UNESCO.

Martin R. (ed.). (2013) Clogging issues associated with managed aquifer recharge methods. IAH Commission on Managing Aquifer Recharge. www.iah.org/recharge/clogging.htm

Megdal S., Dillon P. (eds.). (2015) International Open Access Journal "*Water*" Special Issue "Policy and Economics of Managed Aquifer Recharge and Water Banking". http://www.mdpi.com/journal/water/special_issues/MAR

Miotliński K., Dillon P.J., Pavelic P., Cook P.G., Page D.W., Levett K. (2011) Recovery of injected freshwater to differentiate fracture flow in a low-permeability brackish aquifer. *Journal of Hydrology* 409(1–2), 273–282.

Miotlinski K., Dillon P.J. (2014) Relative recovery of thermal energy and fresh water in aquifer storage and recovery systems. *Groundwater*, 53(6), Nov. 14. Doi: 10.1111/gwat.12286.

NRMMC, EPHC, NHMRC (2009) Australian Guidelines for Water Recycling, Managing Health and Environmental Risks, Volume 2C – Managed Aquifer Recharge. Natural Resource Management Ministerial Council, Environment Protection and Heritage Council National Health and Medical Research Council, Jul. 2009, 237p. http://www.environment.gov.au/resource/national-water-quality-management-strategy-australian-guidelines-water-recycling-managing-1

Nützmann G., Viotti P. and Aagaard P. (eds.). (2006) *Reactive Transport in Soil and Groundwater: Processes and Models.* Springer Science & Business Media.

Page D.W., Vanderzalm J.L., Miotlinski K., Barry K.E., Dillon P.J., Lawrie K., Brodie, R. (2014) Determining treatment requirements for turbid river water to avoid clogging of aquifer storage and recovery wells in siliceous alluvium. *Water Research*, 66, 99–110. http://dx.doi.org/10.1016/j.watres.2014.08.018

Parsons S., Dillon P., Irvine E., Holland G., Kaufman C. (2012) Progress in Managed Aquifer Recharge in Australia. Waterlines Report Series No 73, March 2012, National Water Commission, 107p. http://archive.nwc.gov.au/library/waterlines/73

Pavelic P., Dillon P.J., Simmons C.T. (2006) Multi-scale characterization of a heterogeneous aquifer using an ASR operation. *Ground Water* 44(2):155–164.

Pavelic P., Dillon P.J., Barry K.E., Vanderzalm J.L., Correll R.L., Rinck-Pfeiffer S.M. (2007) Water quality effects on clogging rates during reclaimed water ASR in a carbonate aquifer. *Journal of Hydrology* 334, 1–16.

Pavelic P., Dillon P.J., Mucha M., Nakai T., Barry K.E., Bestland E. (2011) Laboratory assessment of factors affecting soil clogging of soil aquifer treatment systems. *Water Research* 45, 3153–3163.

Peters J. *et al.* (eds.). (1998) Artificial Recharge of Groundwater. Proceedings of TISAR, Amsterdam 1998. A.A. Balkema. http://www.gbv.de/dms/tib-ub-hannover/249900297.pdf

Pyne R.D.G. (2005) Groundwater Recharge and Wells. 2nd Edn. Published by ASR Systems, Florida.

Ray C., Shamrukh M. (eds.). (2011) Riverbank Filtration for Water Security in Desert Countries. Springer and NATO Public Diplomacy Division.

Ross A. (2014) Banking for the future: Prospects for integrated cyclical water management. *Journal of Hydrology* 519, 2493–2500.

Sheng, Zhuping, Zhao, Xuan (2015) Special Issue on Managed Aquifer Recharge: Powerful Management Tool for Meeting Water Resources Challenges. ASCE J Hydrologic Engineering, Vol. 20, No. 3, March 2015. 12 papers http://ascelibrary.org/toc/jhyeff/20/3

SKM (2010) Assessment of Evaporation Losses from the Menindee Lakes using SEBAL Remote Sensing Technology. Report to National Water Commission, April 2010.

Steenbergen F. van, Tuinhof, A. (2010) Managing the Water Buffer for Development and Climate Change Adaptation: Groundwater Recharge, Retention, Reuse and Rainwater Storage. UNESCO and IAH Netherlands Chapter Publication. http://www.bebuffered.com/downloads/3Rbook_2nd_edition_webversion.pdf

Tuinhof A., Heederick J.P. (2003) Management of aquifer recharge and subsurface storage. – making use of our largest reservoir. Netherlands National Committee of International Association of Hydrogeologists, Publication No. 4.

Tuinhof A., van Steenbergen F., Vos P., Tolk L. (2013) Profit from Storage: The costs and benefits of water buffering. Wageningen, The Netherlands: 3R Water Secretariat. http://www.bebuffered.com/downloads/profit-from-storage-reprint-2013_digitalvs.pdf

van den Hoven, T., Kazner C. (eds.). (2009) TECHNEAU: Safe Drinking Water from Source to Tap State of the Art & Perspectives. IWA Publishing.

Vanderzalm J.L. et al. (2009) Water Quality Changes During Aquifer Storage and Recovery. Water Research Foundation. Denver, USA.

WA Department of Water (2010) Operational policy 1.01 – Managed aquifer recharge in Western Australia. http://www.water.wa.gov.au/PublicationStore/96686.pdf

Ward J.D., Simmons C.T., Dillon P.J., Pavelic P. (2009) Integrated assessment of lateral flow, density effects and dispersion in aquifer storage and recovery. *Journal of Hydrology* 370, 83–99.

Ward J., Dillon P. (2011) Robust policy design for managed aquifer recharge. Waterlines Report Series No. 38, January 2011, 28p. National Water Commission. http://archive.nwc.gov.au/library/waterlines/38

Zhao, Xuan, Wang, Weiping (eds.). (2015) Thematic Issue ISMAR8 China. Environmental Earth Sciences, Volume 73, No 12, June 2015: 17 papers http://link.springer.com/journal/12665/73/12/page/1

Chapter 3

Groundwater and urban development in the 21st century – moving from piecemeal development to planned management in developing cities

S. Foster

Visiting Professor of Groundwater Science & Global Water Partnership, Senior Adviser, University College London, London, UK

ABSTRACT

Groundwater is a critical, but often underappreciated, resource for urban water supply, a serious and costly hazard to urban infrastructure, and the 'invisible link' between various facets of the urbanisation process. An overview is presented of the benefits of urban groundwater use, together with some insidious and persistent problems that groundwater can present for urban development. Spontaneous piecemeal approaches invariably mean that 'one person's solution becomes another person's problem' – and there is a strong argument for groundwater considerations to be part of a more holistic approach to urban infrastructure planning and management. However, this is not a simple task because of the widespread vacuum of institutional responsibility and accountability for groundwater in urban areas. This chapter focuses on summarising the current state of urban groundwater management and finding pragmatic solutions to strengthening various facets of urban groundwater governance and management, using examples from Latin America and South Asia.

3.1 SETTING THE GOLBAL CONTEXT

3.1.1 The urbanisation challenge

Urbanisation is a major challenge for water management. Globally, the urban population is expected to grow from around 3.0 to 6.4 billion by 2050, with about 90% of the growth in low-income countries. Urban populations are not only growing but also 'growing-up', which is increasing their water demand disproportionately and generating more waste water per capita. Unless adequately managed these trends are likely to impact negatively on groundwater resources, which must be managed within an integrated framework along with other components of the urban water system, such as surface water, wastewater, and storm water (Foster & Vairavamoorthy, 2013).

3.1.2 Drivers and modes of urban groundwater use

Urbanisation was the predominant global phenomena of the 20th Century and is predicted to continue at increasing rates for the foreseeable future. Groundwater has been a vital source of urban water supply since the very first settlements, when it was

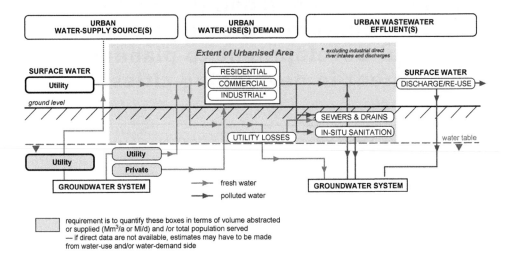

Figure 3.1 Sources and uses of urban water supply and their generation of 'downstream' waste water flows (after Foster & Hirata, 2011).

captured at springheads and by manually-excavated waterwells. Recent decades have seen major growth in urban groundwater use with municipal water utilities, deploying deep waterwell technology and private abstractors in some instances constructing large numbers of low-cost shallow waterwells (Foster *et al.*, 1998).

The principal modes of groundwater use in urban areas are summarised in Figure 3.1. To understand the dynamics of urban water resource accounting it is important:

- to distinguish between utility waterwells constructed within urbanised areas on a piecemeal basis (in response to new demand centres) and protected 'external well fields' or springheads developed as part of a long-term water supply strategy,
- to appreciate that most utilities in the developing world (and some more widely) have high levels of 'unaccounted for' water (often more than 30% and in some cases 50% of the pumped supply), including a major component of physical losses from the distribution system to groundwater together with offtake by illegal connections,
- not to overlook the significance of private self supply from groundwater, for industrial purposes (which is traditional) but also by residential and commercial users, which is not restricted to cities with high yielding aquifers and is a growing phenomenon where the municipal service is inadequate (Foster & Garduño 2002; Foster *et al.*, 1998; Foster *et al.*, 2010b; Gronwall *et al.*, 2010) and often represents a significant proportion of the total water actually received by users, and
- to recognise that any given urban area is in a continuous state of evolution on a time scale of years to decades (Foster & Hirata, 2011).

Comparative examples from Brazilian and Indian cities clearly illustrate these issues and this evolution (Sections 3.1.3 and 3.1.4).

3.1.3   Case Study A: a tale of two Brasilian cities

The following case study is based on Foster & Garduño (2006) and Foster *et al.* (2009). Ribeirão Preto (São Paulo State) is underlain by the major Guarani Aquifer locally partially confined by Serra Geral Basalts. Its metropolitan area had a population of 0.8 million in 2007 (predicted to double by 2040) and is a major industrial centre with important fuel-alcohol distilling, agro-industrial services and a wide variety of manufacturing enterprises. The water supply utility has been able to provide a reliable, moderately-priced service from some 95 utility water wells with an estimated production of 127 Mm^3/a. The total groundwater abstraction (including that of private water wells) has grown from 15 Mm^3/a in 1976 to 186 Mm^3/a in 2007. Contemporary groundwater recharge is exceeded by abstraction. Aquifer water levels across the city have fallen by 30 to 40 m since 1970. This has resulted in substantial increases of operational water-supply costs and water courses becoming influent, which further increases groundwater pollution risk. But groundwater quality from deeper water wells at least has remained excellent. Some pressing issues, however, need to be addressed:

- land use planning for the aquifer recharge zone must be made more compatible with its primary function as a low cost, high quality source of potable municipal water supply,

- risk appraisal for municipal groundwater sources with respect to current urban sanitation measures, industrial activities and agricultural practices to promote appropriate risk management action,

- constraint on demand for groundwater abstraction, since current average water use is very high, and

- consideration of developing some municipal groundwater production capacity from external well fields in the most protected part of the confined aquifer.

Some important advances have already been promoted, including severe constraint on waterwell drilling and/or replacement in the urban area until an integrated resource management plan is in place.

Fortaleza (Ceara State) is a major city with a total population of 3 million growing at rates of 3.5%/a. It is mainly located in a continuous urbanised coastal strip of 2.1 million population at densities of 20–70/ha, underlain by a variable aquifer system of +30 m saturated thickness (comprising Tertiary consolidated sands overlain by Quaternary dune sands). The water service utility supplies some 60–70% of the population from surface water treatment plants with a 'guaranteed yield' of 570 Ml/d, and capacity will increase markedly on completion of major 'canal transfer schemes'. In sharp contrast to Ribeirão Preto, earlier periods of 'near collapse' of mains water-supply during extended droughts (most recently in 1998) led to some 40–60% of the population (mainly in multi-residential properties, together with many commercial users), constructing water wells for direct self supply. A groundwater use survey in 2002–03 recorded 8950 fully equipped water wells (compared to about 1700 in 1980), with 85% being tube wells of +20 m depth) used for domestic water supply. The survey concluded that:

- the sunk capital in private waterwells was at least US$ 19 million and probably US$ 25 million,

- potential groundwater production capacity is about 200 Ml/d (representing 36% of total drought water supply), although actual current use is probably about 80 Ml/d (less than 15% of the total),

- current groundwater abstraction does not tax resources or lead to coastal saline intrusion, since it is balanced by mains water leakage, infiltration from in situ sanitation and drainage soakaway seepage, and

- there are some groundwater quality hazards requiring management, since shallow dugwells often exhibited fecal contamination, and deeper tube wells some 'residual fecal contamination' (15–35 mg NO_3/l & 100–150 mgCl/l) with local 'contaminated hot spots' ($NO_3 > 45$ mg/l & $NH_4 > 2$ mg/l).

Significant groundwater usage arises as a result of domestic properties, avoiding use of mains water supply at prices above the highly subsidised 'social tariff' (US\$ 0.26/m^3). This has major financial implications in terms of loss of revenue from potential water sales, difficulties of increasing average tariffs, and resistance to recovering sewer use charges. A 'sewer use charge' (based on estimated volumes for properties with equipped water wells) has now been levied, but the state water resource secretariat needs to promote a balanced policy, encouraging groundwater utilisation for 'non-sensitive' uses (such as garden watering, laundry processes, car washing, cooling systems, etc.), and reserving high quality mains water for providing a basic potable supply to a larger number of consumers.

3.1.4   Case Study B: a tale of two Indian cities

The following case study is based on Foster & Mandavkar (2008) and Foster & Choudhary (2009). Lucknow City (Uttar Pradesh State) on the Ganges alluvial plain is underlain by Quaternary alluvial sands with 'three productive aquifer horizons' down to 300 m. From 1892 Lucknow City had a limited water supply network based on a small intake works on the Gomti River, but its population grew rapidly from 1 million in 1981 to 2.3 million in 2001, and is projected to reach 4 million by 2020. Conjunctive groundwater use commenced 'incidentally' from 1973, following construction of the first utility water wells, tapping the 'second productive horizon'. Over 300 municipal tube wells have been drilled subsequently (with the more recent well being 200–350 m deep). The available municipal supply by 2005 was 490 Ml/d (240 Ml/d from groundwater and 250 Ml/d from surface water) with the Gomti intake replaced by a Sardhar Canal offtake. In the 1950s, water-table depths were less than 10 m below ground level (bgl), but today they have been widely depressed to 20 m bgl and greater than 30 m bgl in some areas (declining at rates of >1 m/a). However, all utility water wells deliver acceptable raw water quality, but private tube wells tapping the 'first productive horizon' have elevated nitrate. The current operational position is characterised by:

- substantial physical leakage (estimated at 30% overall) reducing deployable supply to about 345 Ml/d,

- source and distribution limitations resulting in a service of about 6 hours/day at low pressure, with individual use of about 100 lpc/d, except in a few areas.

As for most cites on alluvial aquifers there is only a small difference between the economic cost of municipal supply (running costs of US$ $0.12/m^3$) and private tube well water supply (total cost of US$ $0.15–0.3/m^3$). Current charging results in users, paying only US$ $0.04/m^3$. There has been an increasing rate of private water well construction as a 'coping strategy' to secure supply continuity. However, while the long-term availability of local groundwater resources remains good, municipal water engineers tend to favour reducing water well dependence (because of operational complexity) and they prefer major new surface water transfer schemes, rather than more secure conjunctive use based on a new rural protected well field in an area experiencing soil waterlogging.

Aurangabad (Maharashtra State) has grown rapidly and in 2010 had a population of 1.1 million. In the 1960s public water supply was provided from traditional gallery sources providing 5–15 Ml/d, but following the 1972 drought a preferential supply from the Jayakwadi Reservoir (some 45 km away and at 180 m lower elevation) was negotiated. This increased from 28 Ml/d in 1975, 56 Ml/d in 1982, 100 Ml/d in 1992 to 150 Ml/d in 2005, but (aggravated by electrical power shortages) has not been sufficient and municipal utility service levels are poor (widely less than 1-in-24 hours). Thus almost all residential and commercial properties have drilled private bore wells, equipped with small pumps to supplement public supply. Groundwater resources, in terms of resource availability and water well yield, from the underlying weathered Deccan Traps aquifer are limited. Aurangabad has some main collector sewers, but the system is malfunctioning and as a consequence many properties discharge waste water untreated to the extensive pluvial drainage. However, it not clear how this impacts ground water quality. A systematic field survey of water well use in representative electoral wards was conducted in 2008 revealing that:

- residential use increases markedly from February and reaches a maximum (20–50% higher) in May–June,
- household water use is about 0.35 Ml/a (200 lpd/person) a large proportion of which is groundwater,
- the capital cost of water wells is very low (less than US$ 400), but even so private investment totals US$ 1.4–2.2 million per ward with US$ 11–67/month being spent on pumping energy. Thus, private groundwater costs US$ $0.15–0.24/m^3$ as compared to tanker water at around US$ $1.33/m^3$ and the highly subsidised municipal piped supply at US$ $0.03/m^3$ (although actual operational costs are US$ $0.16/m^3$).

The following policy recommendations arise in cities like this in Peninsular India:

- access to groundwater will unquestionably affect the 'willingness to pay' of residential users for improved municipal supply and thus the viability of major new 'imported' water supply schemes,
- in evaluating the benefits and risks of in situ groundwater use, microbiological and chemical quality should be taken more into consideration,
- the use of urban groundwater is in certain ways logical, especially for meeting the demand for sanitary and laundry purposes, where a more expensive treated water-supply may not be justified, and

- municipal authorities should provide fiscal incentives and technical guidance to promote roof and pavement water harvesting for aquifer recharge and pollution risk reduction from waste water.

3.1.5   Scale of groundwater dependency

Most towns and cities located in favorable hydrogeological settings will initially be heavily dependent upon groundwater, although there are rarely sufficient groundwater resources within municipal limits to satisfy the demands of larger growing cities (Foster *et al.*, 1998; Han, 1998; Taniguchi *et al.*, 2009). Available data suggest a broad generic relationship between the size of urban area and groundwater dependency of water utility, albeit with some notable exceptions. Urban centres surrounded by high yielding aquifers (offering potential to expand water supply incrementally with demand) are found to have better utility water services and lower average water production costs. As a result many countries exhibit a high level of dependence on groundwater for urban water supply (e.g. from Denmark and Germany to Brasil, Nigeria, Pakistan, Peru and Vietnam).

Information on groundwater production by urban utilities should be readily available, since it is required for water resource planning, assessing water supply security, and for water-utility 'asset management' inventories. But a recent survey found data difficult to access and frequently deficient. Where private self supply from groundwater is an important component of total urban supply, rough estimates can be made from 'national health surveys', which quantify dependence on 'non-reticulated water wells', and imply huge populations served: 62–82 million in Tropical Africa (with over 40 million in Nigeria alone) and 154 million in seven South Asian countries (Foster & Vairavamoorthy, 2013).

3.1.6   Urban modifications to groundwater cycle

Urbanisation greatly modifies the groundwater cycle with some benefits and numerous threats. The most notable threat is typically a marked impact on groundwater quality resulting in a significant water supply hazard (Figure 3.2a) and unstable urban groundwater regimes, causing an equally serious infrastructure hazard (Figure 3.2b) (Foster *et al.*, 1998; Howard, 2007).

Urbanisation processes interact with groundwater through substantially increasing groundwater recharge rates. Typically the reduction in natural rainfall recharge through land surface compaction is more than compensated for by water mains leakage, infiltrating pluvial drainage, and 'return' of wastewater via in situ sanitation and sewer leakage(Foster *et al.*, 1994). With this increased recharge there is typically an increased contaminant loading from in situ sanitation and to lesser degree sewer leakage. Contamination also occurs due to inadquate storage and handling of 'community' and industrial chemicals, including disposal of liquid effluents and solid wastes. Groundwater systems underlying cities are thus often 'the ultimate sink' for urban pollution. These modifications to groundwater recharge and quality vary systematically with hydrogeological setting. For example, unconfined (oxygenated) aquifers allow free vertical movement of water and rapid transfer of pollutants to the water table. There is also the potential for direct interaction with the built infrastructure. Deeper confined aquifers are overlain by aquicludes/aquitards which greatly impede vertical water movement, often contain anoxic groundwater, and are less prone to pollution but more readily overexploited due to low recharge compared to shallow unconfined systems. The hydrogeological setting also tends to determine the extent of the 'urban groundwater footprint' into the rural hinterland – the precise form being determined by the presence of major cones of pumping depression from external water-utility well fields, natural aquifer recharge and discharge areas, and zones of 'incidental recharge' from urban wastewater irrigation (Foster & Chilton, 2004; Foster & Hirata, 2011).

Very rapid urbanisation, consequent upon large-scale migration of rural population, is leading to both escalating urban land prices from a construction industry boom and to major increases in the extension of informal unplanned slum dwellings. This places a heavy burden on municipal authorities for expansion of water service infrastructure, 'by-passing' land use planning and building regulations, for example, in Bangalore-India and Lusaka-Zambia (Gronwall et al., 2010), provoking groundwater resource degradation and complicating water supply provision.

Later in the evolution of major conurbations, serious impacts to infrastructure can occur due to water table rebound, resulting from abandonment of water well pumping, which can occur with the migration of 'heavy industry' and/or with groundwater pollution fears. This rebound can seriously impact established urban infrastructure, as illustrated by the experience of the past decades in Buenos Aires, Argentina (Foster & Garduno, 2003).

3.2   URBAN GROUNDWATER – RESPONSIBILITY AND ACCOUNTABILITY

3.2.1   Lack of an integrated vision

Groundwater is far more significant in the water supply of developing cities than is commonly appreciated and is also the 'invisible link' between cause and effect in urban water management. Most urban groundwater problems are insidious, persistent, and costly to remediate. They affect everybody, but all too often they are the responsibility of 'nobody'. While many problems are 'predictable', few are actually 'predicted', because of the vacuum of responsibility and accountability (Foster et al., 2010b).

Groundwater resources in and around urban areas are influenced by a complex array of local decisions, which are rarely viewed in an integrated fashion. These include the following (whose jurisdiction is also indicated):

- water well drilling and use authorisation (usually water resource agencies),
- water supply production and distribution (mainly water-service utilities),
- land use change and industrial development (municipal government),
- installing in situ sanitation and handling wastewater (public health departments and water service utilities), and
- handling industrial and community chemicals, and disposing of liquid effluents and solid wastes (environmental authorities and private enterprise).

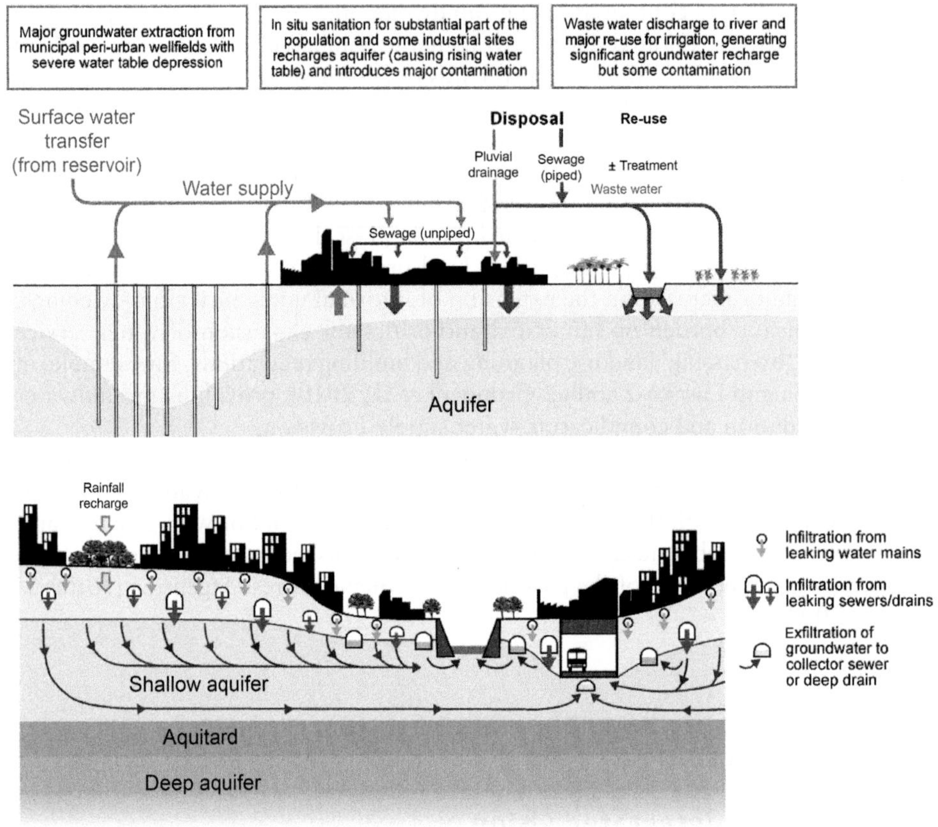

Figure 3.2 Typical modifications of an urban groundwater regime (above) across a typical developing city and (below) in detail in a sewered downtown area.

3.2.2 Filling the institutional vacuum

There is a clear need for groundwater issues to be considered when making decisions on infrastructure planning and investment, but this is not as simple as it might at first appear. Institutional responsibility is often split between various organisations, none of which take the lead (Foster & Hirata, 2011). The reality is that:

- water-resource agencies rarely have the operational capacity or statutory authority necessary to cope with urban development dynamics,
- urban water service utilities in the developing world can be 'resource illiterate',
- urban land and environment departments have little understanding of groundwater,
- river basin committees (where these exist) maybe aware of the need to incorporate groundwater into watershed planning, but rarely have enough knowledge or resources to do more than identify the issue for attention by others.

The dynamics of urban development and its intimate relationship with groundwater merit the formation of a 'standing trans sector urban groundwater consortium'. This would include all major stakeholders and regulatory agencies and be empowered and financed to define and implement a 'priority action plan' (Foster & Vairavamoorthy, 2013). Such consortia should be provided with sound technical diagnosis by an appropriate group of groundwater specialists. The major challenge for such consortia will be promoting differential land management for important recharge areas in the interest of groundwater quality and confronting the impediments resulting from geographically-fragmented land use tenure and environmental control.

3.2.3 Towards effective management planning

Groundwater planning should be an essential component of 'integrated urban water management'. It first requires delineation of 'groundwater management units (or bodies)' (normally centred around urban abstraction), through consideration of the groundwater flow regime (and any major surface water interactions), defined by hydrogeological criteria. Its boundaries should be based on aquifer system hydrodynamics, integrated with local political land divisions, within which the public administration has to work.

Given the evolutionary nature of 'urban groundwater systems' and significant hydrogeological uncertainty in predicting their precise behaviour, an 'adaptive management approach' is strongly recommended. Such an approach needs to be founded on sound information generated by:

- hydrogeological investigation to establish aquifer recharge mechanisms/rates (including man-made sources), the evidence for natural aquifer discharge, the position of saline water interfaces and/or polluted groundwater,
- detailed inventory of current municipal, industrial, and commercial water wells (including the status of their use rights and socio economic profile of their users), economic assessment of the cost of improving interconnectivity within parts of the municipal water supply system to allow areas to be supplied from different sources, and
- evaluation of surface water availability for municipal water supply within various time-frames, the seasonal variability of their yield, and other vulnerabilities.

The core tools for adaptive groundwater management are a calibrated transient numerical aquifer model and an adequate groundwater level and water quality monitoring network. A regularly updated numerical aquifer model can be used to evaluate future groundwater abstraction scenarios (including increased abstraction rates during drought) and thus help to define more robust and sustainable solutions for municipal water supply.

Critically, a groundwater management plan must coordinate with various 'external interfaces' such as sanitation, drainage, and built infrastructure. This requires effective coordination with the corresponding authorities for public health, environment, infrastructure, water supply, and drainage (Foster & Vairavamoorthy, 2013). Similarly, groundwater will also have an intimate relationship with metropolitan and municipal land use planning. A sound groundwater management plan will also need to be in place before large-scale water supply transfers are introduced into an urban area, previously dependent on local groundwater. There are many examples of very costly problems arising where rapid water table rebound and/or increased groundwater pollution has occurred, when groundwater abstraction is significantly reduced (Foster & Chilton, 2004).

3.2.4 Decentralised urban water-service paradigms

Given escalating global rates of urbanisation, future urban water service systems will need to be more decentralised and planned in stages as 'closed-loop' operational cells (servicing populations in the range 10 000–50 000). This type of system can be operated to minimise infrastructure costs, energy use, and water losses, since they reduce the distance between household use and water treatment. They also promote energy efficiency and nutrient recovery. They thus convert current liabilities (like effluent disposal during waste water treatment) into assets (nutrient recovery from waste water treatment), and facilitate local waste water reuse (Vairavamoorthy et al., 2011).

The natural drought resilience and quality protection of deeper aquifers means that they are well suited to be the water supply source for decentralised closed loop water service systems. Since these systems treat waste water as a resource, their installation should substantially reduce the urban subsurface contaminant load from in situ sanitation, reducing a major groundwater pollution threat. Most pathogens from urban waste water are short lived (typical persistence of only one year) once in the subsurface, but risks associated with hazardous synthetic organic compounds (antibiotics and endocrine disruptors, for example) must also be managed. It will be necessary to put more local effort into control of other forms of groundwater contamination (from petrol stations, small-scale motor shops, drycleaning laundries, etc.) to prevent the loss of important water well sources in these local-scale systems

3.3 PRAGMATIC GROUNDWATER MANAGEMENT STRATEGIES

3.3.1 Improving the sustainability of municipal utility use

For the future it will be important that groundwater sources be used on an efficient and sustainable basis for urban water supply. This will require effective demand management measures to constrain inefficient and unnecessary use, and reduction of 'unaccounted for' water, together with groundwater storage being managed strategically to improve long term water supply security (Foster et al., 2011). An integrated approach will be needed, which involves such measures as:

- declaring 'critical areas', where large-scale groundwater abstraction must be constrained or reduced, variously through water well closures and/or specific bans on new or replacement water wells,
- establishing municipal well fields outside cities with capture areas declared as drinking water protection zones (Figure 3.3), with incentives introduced for neighbouring rural municipalities to facilitate their protection,

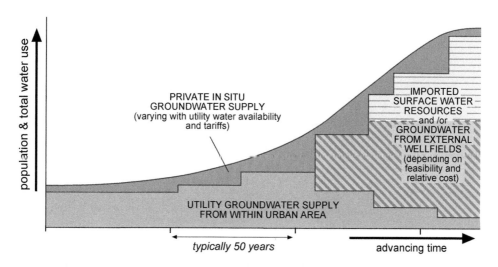

Figure 3.3 Typical evolution of water-supply sources with large-scale urban development (for areas surrounded by high-yielding aquifers) (after Foster & Hirata, 2011).

- importing surface water supplies from distant sources, usually at high associated capital and revenue cost only once demand outstrips local supply (Figure 3.3), and
- developing a strategy that considers a portfolio of options (including mains leakage reduction, storm water capture, and rainwater harvesting), promotes multiple water use (in sequence from higher to lower quality need), and reduces wastage.

This presents not only a challenge in terms of raising the necessary financial investment, but also of overcoming conceptual, institutional, and administrative constraints (Foster *et al.*, 2010b).

3.3.2 Evolution to conjunctive groundwater management

The large groundwater storage of aquifers should be managed conjunctively with surface water (Figure 3.4) to improve water supply security. However, most present conjunctive use in developing nations represents a 'piecemeal coping strategy', for example in Lucknow City, India (Section 3.1.4). There are, however, some good examples of optimised resource use, such as Lima (Peru) and Bangkok (Thailand) where the normal constraints to promoting managed conjunctive use were overcome and the related capital investment mobilised (Buapeng & Foster, 2008; Foster *et al.*, 2010b). However, urban water engineers (pressed by day-to-day problems) more often look for operationally simple setups (such as a single major surface water source and large treatment works), rather than more secure and robust conjunctive solutions. There are also sometimes vested interests in constructing large capital works. A 'resource culture' needs to be fostered within water utilities of developing cities to promote a more balanced view between long-term security (water source characteristics, limitations, and reliabilities) and short-term considerations of operational efficiency and cost.

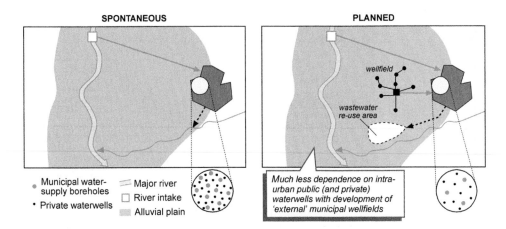

Figure 3.4 Schematic illustration of planned (as opposed to spontaneous) conjunctive use of ground-water and surface water sources for urban water supply (after Foster & Hirata, 2011).

3.3.3 Promoting rational private groundwater use

The capital investment for private water wells is triggered when utility water supplies are inadequate in service continuity, cost, and/or coverage. Private self supply is essentially a 'coping strategy' by households, commerce, and industry (Foster *et al.*, 2010b). Although the 'economy of scale' can be poor, the cost of self supply usually compares favourably with the utility tariff (if based on full cost recovery for surface water supply schemes). For this reason private water well use often continues as a 'cost-reduction tactic' by users to avoid payment of higher tariff levels. The well-researched cases of Aurangabad (India) and Fortaleza (Brazil) clearly reveal this user behaviour and show that residential and commercial/industrial users tend to take water from multiple sources, according to temporal availability and relative cost (Sections 3.1.3 and 3.1.4).

Public administrations need to undertake a broad assessment of private water well use practices in order to formulate a balanced policy on groundwater resource use. Intensive private groundwater use does not necessarily cause serious resource over-exploitation since aquifers are often replenished from water main leakage and seepage from in situ sanitation, but private users are at risk from anthropogenic pollution or natural contamination. Private water well use also reduces demand on municipal resources and is particularly useful for non-sensitive uses such as garden irrigation, cooling systems, and recreational facilities. Private abstraction also guards against the possibility of groundwater table rebound and urban drainage problems should utility abstraction radically reduce.

If there are serious hazards from groundwater pollution or overexploitation the following management actions could be considered (as appropriate to local conditions):

- registering commercial and industrial users, together with residential use for apartment blocks and multi-occupancy estates, charging for abstraction (directly by metering or indirectly by estimated sewer discharge) to constrain use, and

- issuing water-use advice and/or health warnings to private water well operators, and in severe pollution situations declaring sources unsuitable for potable use.

However, when large numbers of more affluent dwellers opt for private water well use, the knock-on effects can be complex, since whilst they free up utility water production capacity to meet the needs of low income neighbourhoods, they simultaneously reduce utility revenue collection and make it more difficult to maintain highly subsidised 'social tariffs' for minimal use.

Private water wells can pose a major challenge for water resource agencies. Modern water well drilling techniques provide rapid access to groundwater for modest capital investment, making it possible for large numbers of users to invest in abstraction infrastructure, which is soon hidden from view, leading to unregulated and illegal abstraction. This situation is counter productive from both the private and public standpoints, but also impedes rational policy design and integrated planning for urban water supply. Private water wells can be regulated by taking advantage of modern technologies (such as geographical positioning and data capture systems) and gaining civil society commitment, using participatory mechanisms with incentives for 'self-registration' and 'self-monitoring'.

An important emerging policy question is: under what circumstances is a complete ban on private residential water wells justified? Historically, private water well use bans or severe constraints were introduced to help control waterborne disease outbreaks. Examples include cholera in 19th century London and in sea ports in the Caribbean in the 1980s. Restrictions were also introduced in Bangkok and Jakarta in the 1990s to limit land subsidence and flood risk. But bans or restrictions usually have high transaction costs and may only be partially successful. In Brazil, abstraction constraints are currently imposed in parts of Ribeirão Preto and São José do Rio Preto (both in São Paulo State) to address problems of local over exploitation, with restrictions applying to all classes of groundwater user. In São Paulo City abstraction controls are in place for zones of proven industrial groundwater contamination, but complete replacement is simply not possible (Foster & Hirata, 2012).

There are some promising examples of regularising private use of urban groundwater. In Bangkok (Figure 3.5) utilising an approach of time limited licensing for all larger multi-residential, industrial, and commercial groundwater abstractors was adopted to constrain private water well use in critical areas. This was coupled with a progressive charging plan which has successfully stabilised groundwater levels and curtailed serious land subsidence (Buapeng & Foster, 2008). In Recife and Fortaleza (Brazil), municipal utilities argued for levying a volumetric water charge on private waterwell users who make use of mains sewerage. A comprehensive inventory of private water wells on multi-residential, commercial, and industrial properties was drawnup and charges made based on sewer use by type/size of property or by metering private water well use (Foster et al., 2010b).

3.3.4   Mitigating in situ sanitation pollution hazards

In developing nations, urban in situ waste water disposal is extensive and presents a significant groundwater hazard, which needs to be recognised and managed (Foster et al., 1998; Barrett & Howard, 2002). The threat posed by waste water disposal is usually

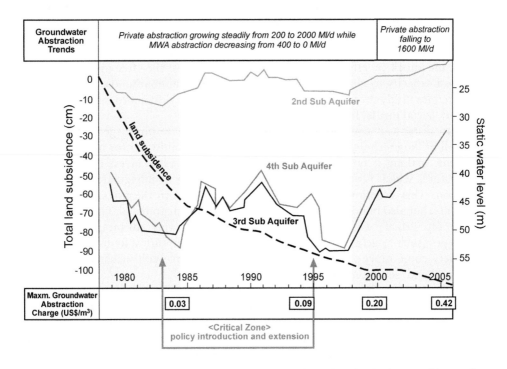

Figure 3.5 Evolution of groundwater abstraction, water-levels, and land subsidence in Metropolitan Bangkok with improving resource management (after Buapeng & Foster, 2008).

more widespread than that posed by inadequate handling of industrial chemicals and disposal of industrial effluents. In most aquifer types (except the most vulnerable and shallow) there will be sufficient natural groundwater protection to eliminate faecal pathogens in percolating waste water. Hazards increase markedly with substandard water well construction and/or informal or illegal sanitation and waste disposal practices, for example in numerous cities of tropical Africa (Foster, 2009).

However, troublesome levels of nitrogen compounds (usually nitrate, sometimes ammonium) and dissolved organic carbon will also typically evolve in vulnerable aquifers to varying degree. This will be a function of population density, served by in situ sanitation and aquifer properties. Such pollution can penetrate to considerable depths in aquifers and persist after the contamination source is removed by installation of main sewerage or other alternative sanitation (Foster *et al.*, 2011). The most cost effective way of dealing with this type of problem in municipal water supplies is by dilution through mixing, which requires a secure and stable source of high quality supply, such as that produced from protected 'external well fields', for example as is practised in Natal-Brasil (Foster *et al.*, 2010). Tertiary treatment (to drinking water standards) is usually only possible in more affluent countries.

A more integrated approach to urban water supply, mains sewerage provision, and urban land use is required to avoid persistent and costly problems, especially where local aquifers are providing the municipal water supply. Public administrations and water service providers can employ a number of simple measures to improve groundwater sustainability (Drangert & Cronin, 2004; Foster *et al.*, 2010b). These include:

- prioritising recently urbanised districts for sewer coverage to protect good quality groundwater and/or limiting the density of new urbanisation (served by in situ sanitation) to contain groundwater nitrate contamination,
- take advantage of parkland or low density housing areas to site utility water wells,
- establishing groundwater source protection zones around all utility water wells that are favourably located, and
- involving residents in waste water quality improvement by seeking cooperation on not discarding unwanted chemical products to toilets or sinks and avoiding the use of particularly hazardous chemicals in the community.

Much more effort is needed to change attitudes towards waste water reuse and associated energy costs and nutrient recovery, which can contribute positively to urban groundwater management. New technologies that promote waste water as a resource need to be tailored to conditions in low income countries, including low cost membrane systems, and hybrid natural and constructed wetlands. Another promising technology is the so-called ecosanitation (which separates urine from faeces and recovers both for reuse), thus greatly reducing subsurface contaminant load. Although this has limitations since large-scale retroinstallation in existing dwellings is not straightforward and it is less suitable for cultural groups who use water for anal cleansing (Foster & Vairavamoorthy, 2013).

3.3.5 Addressing industrial pollution hazards

Where there is significant industrial activity interspersed with public utility and private domestic water wells, it is essential to carry out groundwater pollution surveys and risk assessments. Fuel storage facilities, chemical plants, paint factories, metallic and electronic industries, dry-cleaning establishments, leather tanneries, timber treatment, and waste tips can all discharge mobile, persistent, and toxic chemicals with the potential to contaminate groundwater and thus need to be closely monitored. Intensity of subsurface contamination is not necessarily a function of the size of industrial activity or facility. Often small, widely distributed enterprises use considerable quantities of toxic chemicals and pose a major threat since they operate outside the formal registers and environmental controls.

Groundwater pollution surveys and risk assessments should be commissioned by the public health, environmental, or water resource agencies, in close liaison with water service utilities, using recommended guidelines and protocols (Foster *et al.*, 2002). A typical survey would involve the following steps:

- a systematic survey of existing and past industrial activity to assess the probability of different pollutant types contributing to subsurface contaminant load,
- a groundwater pollution hazard assessment, considering the interaction between the subsurface contaminant load and local aquifer pollution vulnerability,

- detailed groundwater sampling and analysis programme with the analytical parameters being guided by the above survey.

The results of such scientific survey and assessment work should guide policy by:

- introducing pollution control measures, including better constraints on handling and disposal of industrial effluents to reduce groundwater pollution risk,
- increasing quality surveillance for selected utility water wells and/or progressive investment to replace water wells considered at greatest risk of serious pollution,
- advising and warning private domestic water well users of potential pollution risks, imposing use constraints, and in extreme cases forced closure of waterwells, and
- designing a long-term focused groundwater monitoring programme to improve water quality surveillance and security.

3.3.6   Spatial planning of downstream waste water reuse

Many developing cities have to invest in expanding mains sewerage and in one sense urban waste water is the only 'natural resource' whose global availability is steadily increasing. Waste water reuse within and downstream of cities for agricultural and amenity irrigation often results in major recharge to alluvial aquifers because of a general tendency to over irrigate. This 'accidental' groundwater recharge often ends up being the predominant reuse in volumetric terms (Foster & Chilton, 2004). Urban waste water must be regarded as both a useful resource but, also a pollution hazard, because its nitrogen content generally exceeds crop requirements and it contains elevated dissolved organic carbon (DOC) concentrations (leading to a trihalomethane hazard on water supply disinfection and/or the possible presence of hazardous synthetic organic compounds in water wells).

The impact of waste water irrigation reuse on groundwater quality can be reduced and managed by a combination of measures (Foster & Chilton, 2004):

- foremost taking a proactive approach to spatial planning controls over waste water irrigation to avoid municipal water well protection areas,
- reducing ingress of saline and toxic water to main sewerage systems,
- urging constraints on the use of shallow private domestic water wells in waste water irrigation areas, but encouraging pumping from shallow aquifers for irrigation,
- improving waste water treatment and reducing overirrigation,
- increasing the intake depths and sanitary sealing of utility water wells, and
- intensifying groundwater monitoring for pathogens and synthetic organics.

3.4   CONCLUSION

Groundwater is the 'invisible link' between various facets of the urbanisation process. Thus, if the full benefits of urban groundwater use are to be realised in finding solutions to the major water supply demands of innumerable fast growing urban centres in the developing world, the much more holistic approach to groundwater within urban infrastructure planning and management described in this paper will need to be taken. This in turn will require considerable pragmatism to developing appropriate institutional arrangements for such a holistic approach to be adopted.

ACKNOWLEDGEMENTS

The author wishes to acknowledge the contribution of all those associated with the World Bank GW-MATE initiative during 2001–11 for developing much of the field experience on which this overview is based. That programme was coordinated sequentially by the late John Briscoe, Karin Kemper, and Catherine Tovey in part to assist various national government agencies in developing nations to formulate more appropriate policies on urban groundwater use. The GWP-TEC Chair, Mohamed Ait-Kadi, is thanked for his personal interest in the topic. The author has also benefited down-the-years from discussion on urban groundwater issues in a range of geographical contexts with Ken Howard, Ricardo Hirata, Brian Morris, Shrikant Limaye, Martin Mulenga, Jochen Eckart, Krishna Khatri, Makoto Taniguchi, Binaya-Raj Shivakoti, Roelof Stuurman, and Radu Gogu.

REFERENCES

Barrett M.H., Howard A.G. (2002) *Urban groundwater and sanitation – developed and developing countries*. In: *Current Problems of Hydrogeology in Urban Areas*. Kluwer (Dordecht) 39–56.

Buapeng S., Foster S. (2008) Controlling groundwater abstraction and related environmental degradation in Metropolitan Bangkok–Thailand. GW·MATE Case Profile Collection 20. World Bank (Washington DC). www.worldbank.org/gwmate.

Drangert J.O., Cronin A.A. (2004) Use and abuse of the urban groundwater resource: implications for a new management strategy. *Hydrogeology Journal* 12, 94–102.

Foster S.S.D., Morris B.L., Lawrence A.R. (1994) *Effects of urbanization on groundwater recharge*. In: *Groundwater Problems in Urban Areas*. Institution of Civil Engineers (London), 43–63.

Foster S., Lawrence A., Morris B. (1998) *Groundwater in urban development – assessing management needs and formulating policy strategies*. World Bank Technical Paper 390 (Washington DC).

Foster S., Hirata R., D'Elia M., Paris M. (2002) *Groundwater quality protection – a guide for water-service companies, municipal authorities and environment agencies*. World Bank GW-MATE Publication (Washington DC).

Foster S.S.D., Chilton P.J. (2004) Downstream of downtown – urban waste water as groundwater recharge. *Hydrogeology Journal*, 12, 115–120.

Foster S. (2009) Urban water-supply security–making the best use of groundwater to meet demands of expanding population under climate change. In: *Groundwater & Climate in Africa*. IAHS Publication 12, 115–120.

Foster S., Garduño H. (2002) Actual and potential regulatory issues relating to groundwater use in Gran Asuncion-Paraguay. GW·MATE Case Profile Collection 3. World Bank (Washington DC). www.worldbank.org/gwmate

Foster S., Garduño H. (2003) Mitigation of groundwater drainage problems in the Buenos Aires Conurbation-Argentina–technical and institutional way forward. GW·MATE Case Profile Collection 4. World Bank (Washington DC). www.worldbank.org/gwmate.

Foster S., Garduño H. (2006) Groundwater use in Metropolitan Fortaleza, Brazil. GW·MATE Case Profile Collection 14. World Bank (Washington DC). www.worldbank.org/gwmate

Foster S., Mandavkar Y. (2008) Groundwater use in Aurangabad – a survey and analysis of social significance and policy implications for a medium-sized Indian city. GW·MATE Case Profile Collection 21. World Bank (Washington DC). www.worldbank.org/gwmate

Foster S., Choudhary N.K. (2009) Lucknow City–India: groundwater resource use and strategic planning needs. GW·MATE Case Profile Collection 23. World Bank (Washington DC). www.worldbank.org/gwmate

Foster S., Hirata R. (2011) Groundwater use for urban development – enhancing benefits and reducing risks. SIWI On the Water Front–Selections from 2011 World Water Week in Stockholm 21–29.

Foster S., Hirata R., Vidal A., Schmidt G., Garduño H. (2009) The Guarani Aquifer initiative – towards realistic groundwater management in a transboundary context. GW·MATE Case Profile Collection 9. World Bank (Washington DC). www.worldbank.org/gwmate

Foster S., Vairavamoorthy K. (2013) Urban groundwater – policies and institutions for integrated management. GWP Perspectives Paper (Stockholm) 18pp.

Foster S., van Steenbergen F., Zuleta J., Garduño H. (2010a) Conjunctive use of groundwater and surface water – from spontaneous coping strategy to adaptive resource management. GW·MATE Strategic Overview Series 2. World Bank (Washington DC) www.worldbank.org/gwmate.

Foster S., Hirata R., Garduño H. (2010b) Urban groundwater use policy–balancing the benefits and risks in developing nations. GW·MATE Strategic Overview Series 3. World Bank (Washington DC) www.worldbank.org/gwmate.

Foster S., Hirata R., Howard K.W.F. (2011) Groundwater use in developing cities: policy issues arising from current trends. *Hydrogeology Journal* 19, 271–274.

Foster S., Hirata R. (2012) Sustainable groundwater use for developing country urban populations – lessons from Brazil. *Water* 21(April 2012), 44–48.

Gronwall J.T., Mulenga M., McGranahan G. (2010) Groundwater and poor urban dwellers – a review with case studies of Bangalore and Lusaka. IIED Human Settlements Working Paper 25. International Institute for Environment & Development (London).

Han Z. (1998) Groundwater for urban water supplies in northern China – an overview. *Hydrogeology Journal* 6, 416–420.

Howard K.W.F. (2007) Urban groundwater – meeting the challenge. IAH Selected Paper Series 8. Taylor & Francis (Oxford).

Taniguchi M., Dausman A., Howard K., Polemio M., Lakshmana T. (2009) Trends and sustainability of groundwater in highly-stressed aquifers. IAHS Publication 329.

Vairavamoorthy K., Tsegaye S., Eckart J. (2011) Urban water management in cities of the future: emerging areas in developing countries. SIWI On the Water Front – Selections from 2011 World Water Week in Stockholm: 42–48.

Chapter 4

Transitioning to sustainable groundwater management in Maiduguri, Nigeria

A. Bakari

The Department of Earth Sciences, Federal University, Birnin-Kebbi, Kebbi State, Nigeria

ABSTRACT

Groundwater is the most reliable water supply source for domestic, agricultural, and industrial use in Nigeria. However, it is under increasing pressure from above ground anthropogenic activities related to uncontrolled urbanisation, incessant waste disposal, and poor land use management. The resultant effect on water users is intolerable and the purely technical solutions of the 20th century have failed to remedy the increasing contamination of this precious resource. This study evaluates the existing problems of groundwater management and establishes a platform for engagement of the various stakeholders whose involvement is required to address the problem. The steps taken to identify and engage the stakeholders responsible for and affected by problems of groundwater contamination are described and discussed. The chapter assesses the existing approaches to groundwater management, the causes of groundwater contamination, stakeholde's capacities and concerns, and provides sustainable solutions that will ensure the transition from the current to a more integrated and viable system. Sustainable strategies that can be applied in many developing countries are recommended. The outcome will be a useful part of the solution for water managers, policy, and decision makers in implementing a sustainable groundwater management systems both in Nigeria and in other rapidly developing countries.

4.1 INTRODUCTION

Groundwater plays a vital role in the development of urban and rural areas in Nigeria. In a recent report it was estimated that the groundwater potential of aquifers in Africa are 100 times the amount found on the surface (MacDonald *et al.*, 2011; McGrath, 2012). Out of the current population of Nigeria of about 170 million, more than half depend directly on this natural resource for their daily water needs. With the rapid population growth of about 2.9% per annum, the demand for water has progressively increased over the last three decades. The provision of safe drinking water has deteriorated, for example, access in urban areas fell from 79 per cent to 75 per cent during 2002. This is largely due to poor management, inadequate technical capabilities and human capacities, insufficient investment, and funding. Others are lack of stakeholder participation in the management of groundwater resources and the fragmented nature of institutions responsible for water management (Jacobsen *et al.*, 2012).

Rapid population growth and uncontrolled urbanisation further aggravate the increasing magnitude and distribution of above ground human activities that potentially affect the quality and quantity of the underlying groundwater (Foster *et al.*, 1998). Uncontrolled urbanisation, dense population concentrations, and ever increasing human activities all severely affect groundwater quality. This is especially the case in developing countries like Nigeria, where the urban expansion is not normally guided by regulations (Chilton, 1999; Wakida & Lerner, 2005; Naik *et al.*, 2008; Putra, 2008; Eni *et al.*, 2011).

The problems pose a significant threat to water quality in the upper unconfined aquifer system of the Chad Basin around Maiduguri in north-eastern Nigeria. This aquifer is a major water supply source for the city, with more than 80% of the residents obtaining their domestic water supplies directly from it (Bunu, 1999). The aquifer is hydraulically connected to the Ngadda River, which drains the city and is highly polluted (Isiorho and Matisoff, 1990). This river–groundwater system is threatened by the impact of on-site sanitation systems (pit latrines and dumpsites) and other non-point sources of contamination across the city (Bakari, 2014).

This negative impact is significant in many areas of Maiduguri metropolis where human, residential, and commercial wastes are indiscriminately disposed of. Also, the hydraulic connectivity, between the river and the upper aquifer, serves as a pathway for groundwater contamination due to inflow of poor quality river water into the aquifer. As a consequence, it poses unacceptable health risks to the local population, most especially on the urban poor who largely depend on groundwater.

Thus, a change (or transition) to a better managed groundwater resource is critical. This transitional change requires identifying, engaging, and empowering the relevant stakeholders in addressing groundwater management issues. It is a continuous process of (radical) change in a society which involves coevolution, of institutional, technological, socio-cultural, and ecological developments at different scales and levels (Rotmans *et al.*, 2000).

The objective of this chapter is to evaluate the problems of groundwater contamination in Nigeria, using a methodology for engagement of the various stakeholders in addressing the situation, which can be used in other similar international cases. The paper explored the current approach to groundwater management, causes of groundwater contamination, stakeholder's capacities, and concerns in providing sustainable solutions that will ensure the transition to a better integrated, more sustainable system.

4.2 STUDY AREA

4.2.1 Description of study area

Maiduguri serves as a gateway to the Sahel region of West Africa. The city is the capital of Borno State in north-eastern Nigeria (Figure 4.1). The area lies on a vast sedimentary basin with an average elevation of 300 m above sea level. The climate is semi-arid with three distinct seasons:

- a long hot, dry season from April to May. Daytime temperatures are in the range of 36–40°C and the night time temperatures fall to 11–18°C.

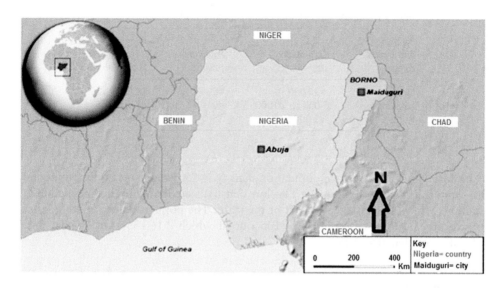

Figure 4.1   Map of Nigeria, showing Maiduguri, Borno state (modified from VOA maps, 2014). Note the location of the cross section in Figure 4.2.

- a short rainy season from May to September with a daily minimum temperature of 24°C and a maximum of 34°C with relative humidity of 40–65%. Annual rainfall is typically between 560 to 600 mm per year (average 580 mm per year).
- a the cold season runs from October to March when temperatures fall to about 20°C (average) with the relative humidity averaging 48 to 49 percent. Potential evaporation averages 1200 mm per year (Jaekel, 1984; Bakari, 2014).

The River Ngadda, which dominates the surface water drainage of the area, flows through Maiduguri. The Ngadda River, like other rivers in the Chad Basin is ephemeral and flows from August to February, with peak discharge typically in September. The Ngadda River cuts through Bama Ridge at Maiduguri in a well-developed water gap. Below the gap, it flows through a system of braided channels and deltaic deposits. Farther north, the Ngadda River gradually loses its identity as it fingers out on a marshy plain. Upstream from Maiduguri, the Ngadda passes through Lakes Yare and Alau, both of which are perennial. The outlet from Lake Alau joins the bypass channel south of Maiduguri (Bakari, 2014).

4.2.2   Population demographics

Nigeria is the most populous country in Africa with an 170 million people and an average density of about 135 persons per km². The population has been estimated as growing at an average of 2.9% per annum, which is putting ever increasing pressure on the aquifer systems. The best estimate of the population distribution of the country (Table 4.1) indicates that the urban areas (major cities) have the greatest population followed by towns and rural areas (NPC, 2006).

Table 4.1 Population distribution by category (NPC 2006).

Population distribution category	Community size	Population (million)	% of total
Urban	>20 000	45	38
Small towns	5 000 to 20 000	40	33
Rural	<5 000	35	29

In Borno state, the average population density is 55 persons per sq km, only one third that of the rest of the country. This low density is attributed to the harsh climate conditions which affect a greater part of the state. The population of the study area (Borno state) in 2006 stood at 4 171 105 out of which 2 163 358 were males and 2 007 746 were females (NPC, 2006).

4.2.3 Geology and hydrogeology

The geology of Nigeria is dominated by three major components: the Precambrian Crystalline Basement Complex (~600 Ma), the Jurassic younger granites (200–145 Ma), and the Cretaceous–Tertiary sedimentary series (≤145 Ma). The Basement Complex rocks are primarily metamorphic and igneous-volcanic in origin.

The main sedimentary basins are the Chad, Sokoto, Bida/Nupe, and Anambra Basins, with the other being the Benue Trough, Benin, and Niger Delta Basins. Generally, the sedimentary sequence in these basins is: a basal unit of non-marine interbedded sandstones, siltstones, and mudstones; a middle unit of marine shales and limestones intercalated with sandstones and siltstones; and an upper unit of sandstone (Obaje, 2009).

The Chad Basin is a structural depression which originated in the early Tertiary period and has been a locus of subsidence and sedimentation rather than erosion ever since. The Chad Basin was a tectonic cross-point between an NE–SW trending 'Tibesti-Cameroon Trough' and the NW–SE trending 'Air-Chad Trough' in which over 3600 m of sediments have been deposited. The Crystalline Basement Complex outcrops in the eastern, south-eastern, south-western, and the northern rims of the basin. The basin appears to be a horst and graben structure, but this has not been confirmed (Oteze & Fayose, 1988).

The Chad Basin has two sub-basins, the Borno Sub-basin and the Cameroun-Chad Sub-basins. The stratigraphy of the Chad Basin (in particular the Borno Sub-basin) shows a depositional sequence from top to bottom, which includes the younger Quaternary sediments, the Plio-Pleistocene Chad Formation, the Turonian–Maastrichtian Fika shale, the late Cretaceous Gongila formation, and the Albian Bima Formation (Maduabuchi et al., 2006). The Bima Sandstone forms the deeper part of the aquifer series and rests unconformably on the basement complex rocks. Its thickness ranges from 300 to 2000 m and depth between 2700 and 4600 m (Obaje, 2009). Barber & Jones (1960), revealed that the Chad Formation reaches a thickness of at least 548 m at Maiduguri, in the central part of the basin, but the thickness may reach 600

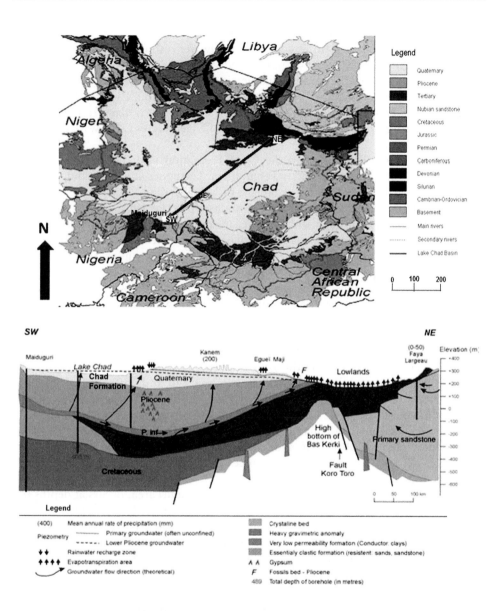

Figure 4.2 Cross section of (SW–NE) the multi-layered aquifer system of the Chad Basin (Borno Sub-basin). Modified from (Schneider & Wolff 1992).

to 700 m elsewhere (Offodile, 1992). The Plio-Pleistocene Chad Formation and the Quaternary sediments are the main sources of groundwater supply in the Maiduguri area (Figure 4.2).

The Chad Formation dips gently east and north-east towards Lake Chad in conformity with the slope of the land surface. Except for a belt of alluvial deposits around the edge of the basin, the formation is of lacustrine origin and consists of thick beds of clay intercalated with irregular beds of sand, silt, and sandy clay (Barber and Jones, 1960; Miller *et al.*, 1968; Odada *et al.*, 2006; Adelana, 2006).

Nigeria is endowed with an abundant groundwater resource, that is far greater than the available surface water resources (Nwankwoala, 2011). The amount of renewable groundwater (based on recharge) is estimated to be 224×10^{12} L/year (Hanidu, 1990; Nwankwoala, 2011). In an earlier investigation, Rijswlk (1981) estimated groundwater storage to be 6×10 km³ (6×10^{18} m³). The potential groundwater resources are estimated in the Sedimentary Basins of Nigeria report (FMWRRD, 1995) to be 5.93×10^9 m³.

Barber & Jones (1960) divided the Chad Formation into three water-bearing zones designated upper, middle, and lower aquifers. The upper aquifer is composed of Quaternary alluvial fan and deltaic sediments of Lake Margin origin. It is composed of interbedded sands, clays, silts, and discontinuous sandy clay lenses. Hydraulically it ranges from unconfined through semi-confined to confined (Maduabuchi *et al.*, 2006). It extends from the surface to an average depth of 60 m, but in some localities has been found at depths as deep as 180 m. The transmissivity of this aquifer system ranges from 0.6 to 8.3 m²/day and the aquifer yield in Maiduguri is between 2.5 to 30 l/s (Akujieze *et al.*, 2003). This aquifer is mainly used for domestic water supply (hand dug wells and shallow wells), which supports vegetable growing and livestock watering (Maduabuchi, 2006).

The middle aquifer is the most extensive aquifer in the Chad Basin (Borno Sub-basin) and it underlies at least 51 800 km² of north-eastern Nigeria. A clay layer, 60 to 300 m thick, confines the water in this zone and separates it from the overlying upper zone. The middle aquifer occurs at a depth between 240 and 380 m below ground level, and consists of 10–40 m-thick sand beds with interbedded clays and diatomites of early Pliocene in age. The sand fractions consist of medium- to coarse-grained quartz, feldspar, mica, and hematite. The average transmissivity value of this zone is 360 m²/day (Obaje, 2009). The aquifer yield of this zone is between 24 to 35 l/s (Akujieze *et al.*, 2003) with a recharge source from infiltration of rainwater via the upper zone.

The lower zone, presently known only in the Maiduguri area, occurs at depths of 423 to 510 m below ground level and consists of about 76 to 200 m of interbedded clay, sandy clay, and, sand (Akujieze *et al.*, 2003). In some parts of the basin, this aquifer is artesian in nature, but it is not utilised for domestic water supply in the greater part of the Chad Basin. Its yield is between 10 to 35 l/s (Akujieze *et al.*, 2003) and its recharge source is unknown.

4.2.4 Institutional arrangement for the management of water resources in Nigeria

In Nigeria, all three tiers of government are involved in the management of water resources. This is because the management and development of water resources is in the concurrent legislative list of the Constitution of the Federal Republic of Nigeria. The general management of water resources is the exclusive responsibility of the Federal Government of Nigeria (FGN) through the Federal Ministry of Water Resources (FMWR), supervised by the Minister for Water Resources with the responsibility to implement all national policies, Federal laws, and regulations relating to water resources management and development.

The institution responsible at a Federal level is the FMWR, which also administers the River Basin Development Authorities. At the State level, the management of water resources, is carried out by the various (36) state ministries of Water, Agriculture, and Natural Resources. At the local level, rural water supplies and sanitation services are carried out by the various local council administrations (774 across Nigeria).

The FMWR was created in 1976 and is liable for formulating and coordinating national water policies, management of water resources, together with allocation of water between states, and approving development projects. The (12) River Basin Development Authorities (RBDA) were created in 1976 for planning and development of water resources, irrigation work, and the collection of hydrological, hydrogeological, and meteorological data. Their main involvement in potable water supply has been the provision of multipurpose dams and the supply of water in bulk, mostly to urban water systems (Goldface-Irokalibe, 2002).

At the state level, the duty for potable water supply was usually vested on the State Water Agencies (SWAs) (36 in number). Each SWA manages water supply facilities within its respective area of jurisdiction. The SWA are answerable to their state governments, generally through a commissioner of the State Ministry of Water Resources (SMWR).

At the local level, all the 774 Local Government Authorities or Councils (LGAs or LGCs), are directly responsible for the provision of rural water supplies and sanitation facilities in their areas. However, only a few of these organisations have the resources and expertise to address these problems. At present, only a handful of LGAs have rural water supply divisions, which are be able to construct small water systems, such as open wells and small impounds of surface water.

4.2.5 Approaches to water management in study area

There is currently no well-defined groundwater management approach in the case study area. All three levels of government are involved in the development and supply of water and water-related services in an ad hoc fashion. The National Policy on Water and Sanitation (2000) mandates the various tiers of government in the country to supply water resources to rural and urban areas. For rural water supply, the FGN's involvement is 50%, State (25%), Local Government Areas (LGAs) (20%), and Community (5%). For the urban areas, the FGN has 30% responsibility, State government has 60%, while 10% is reserved for the LGA (Goldface-Irokalibe, 2002).

The Chad Basin Development Authority (CBDA) is one of the river basin development authorities (RBDAs) in Nigeria under the aegis of FMWR. It was set up specifically to take charge of the water resource development in an integrated manner. The state and local governments are tasked with the responsibility of supplying the inhabitants of the urban and rural areas, respectively. This arrangement was initially ineffective and has remained so, resulting in no clear roles and responsibilities and subsequently conflicting regimes.

4.2.5.1 Availability and access to water resources in the study area

Groundwater resources have been and still remain the principal source of water supply for domestic, livestock, agricultural, and industrial use. The largest demand is from

(a)   (b)

Figure 4.3(a and b) Lake Alau dam reservoir and the dam outlet.

domestic and agricultural needs that account for more than 80% (Bunu, 1999). The high dependence on groundwater is because of its reliability as a source of water supply across the state. Also, it provides a buffer against climatic variability. Its quality is often good and infrastructure is affordable to low income individuals and communities (McDonald & Adelana, 2008). Currently, groundwater is developed in the state through development of private boreholes by individuals, organisations, and commercial ventures. This constitutes over 80% of the total water supply in Maiduguri (UN, 1988; BGS, 2003). Despite a high degree of groundwater availability, access to water supply in urban areas fell from 79% to 75% (Jacobsen *et al.*, 2012). This is because expansion of water utilities failed to keep up with the pace of population growth (WHO/UNICEF, 2000).

A surface water supply scheme provides piped water to the city via a network of underground pipes pumped from the Maiduguri water treatment plant. The plant is operated by Borno State Ministry of Urban and Rural Water Supply. This water treatment plant was designed to treat surface water pumped from Lake Alau, a 162 million m³ reservoir (Figure 4.3). The lake is south-east of Maiduguri and 14 km away from the city along the Maiduguri-Bama road. It has the capacity to treat about 67 000 m³ of raw water daily, which is equivalent to 15 million gallons per day. The raw water is pumped from the Lake Alau dam via a 12 km network of 800 mm (D.I.) pipe to the plant, where it undergoes chemical treatment, clarification, and disinfection before it is distributed to the municipality.

However, the operational capacity of the plant has declined to its lowest level due to the problems of excessive evaporation rates and the infiltration of the water into the unconsolidated sands below. Others are low level of commitment and neglect over the years. Successive governments failed to implement strategies that will boost the capacity of the water works.

4.2.6   Groundwater management problems in the study area

Major environmental problems such as solid and liquid waste disposal by residents and local businesses, proliferation of pit latrines, and other non-point pollution sources related to anthropogenic activities continually pose a significant threat to the potability of the shallow aquifer system. These environmental problems are aggravated by rapid population growth and uncontrolled urbanisation.

Another problem intrinsic to this area is stakeholder exclusion in planning and management of groundwater resources at the community and state levels. Poor coordination among the various components of the water management system is a major constraint in achieving sustainability. Currently, this factor affects the effectiveness of the local water management institutions. The singular approaches adopted by the state water agency have undermined the utilisation and management of groundwater resources.

4.3   METHODOLOGY

A stakeholder analysis outlined in Bakari *et al.* (2014) was adopted to identify all the relevant stakeholders whose input is required to address the environmental problems of the case study area. The analytical tools used in categorising the stakeholders include those using levels of interest and impact (Hare & Pahl-Wostl, 2002) and legitimacy and influence (Mitchell *et al.*, 1997).

In this respect, a total of six stakeholder groups consisting of 22 individual groups and organisations (Table 4.2), Civil Society Organisations (CSOs), non-Governmental Organisations (NGOs), Government Ministries/Agencies, and a research institution were included. Others relevant groups were professional organisations (hydrogeologists and engineers etc.), traditional rulers and local politicians. Representatives of the various stakeholder groups (key stakeholders) were engaged via interviews, focus group discussions, and stakeholder meetings (McNamara, 1999; Morgan, 1997; Patten, 2001). A summary of each of these groups and the survey method utilised for each group is given below. The results of these interviews, discussions, and surveys are detailed in Bakari *et al.* (2014).

Semi-structured interviews with open-ended questions were carried out to explore the current approaches of groundwater management in Maiduguri. Key issues such

Table 4.2   Summary of the various stakeholder groups engaged in the study.

Organisation type	Number of groups
Government ministries/agencies	10
Water user groups	4
Professional organisations	3
Civil society organisations	3
NGO	1
Research institution	1

as groundwater contamination and the potentials for participatory groundwater management in the study area were explored. A total of twelve representatives (strategic
stakeholders) were interviewed between January and March 2013.

Different focus group discussions were held across the study area. In total, there
were 52 individuals, 40 males and 12 females drawn from the residents and water user
groups. Each focus group comprised a dozen residents from these local communities.
The sessions formed open discussions on groundwater issues, such as, knowledge of
levels of groundwater pollution, common causes of contamination, the type of wastes
generated, and disposal methods.

Stratified random sampling outlined (Patten, 2001) was used to identify the various
households for the study. A respondent was identified in every third house in the
selected communities. Participants were selected on sub-divisions of the study area.
The communities were selected on their socioeconomic and demographic background.
A total of 600 household questionnaires were distributed, with an 81% response rate
being achieved.

4.4 DISCUSSION

A discussion of the issues identified, building on the findings of Bakari *et al.* (2014),
shows that environmental problems, impacting negatively on groundwater resources
are widespread in the study area. So, accordingly most interviewees are familiar with
these issues, at least at a basic level. However, in a few instances some of the interviewees failed to give convincing accounts of these issues. The interviewees from
academia, Ministries of Water, Environment, and Health were the most knowledgeable, likely related to their high level of education and professional involvement in
dealing with environmental issues in their respective roles. Despite the differences in
their understanding, all interviewees were keen to be involved in addressing the environmental problems. This is probably because they are in a position of authority, hence
they see it as a vested responsibility as far as their organisations are concerned.

Conversely, awareness about groundwater contamination is very limited in the
general population focus group category. Participants in this category are typically
individuals with little relevant education such as farmers, local business owners, and
traders that constitute the bulk of the urban, less affluent population. Similarly, the
household survey revealed that most of the respondents are not knowledgeable about
groundwater contamination. With more than 87% (n = 288) of the households unfa-
miliar, only a minority (12.2%) of the respondents are informed about this issue.
Survey results clearly indicated a low level of environmental awareness among the
general population.

The majority of those interviewed from a relatively highly educated background
were worried that consuming contaminated groundwater can be very harmful to
human health. The respondents from the relatively poorly educated background typically showed little interest in issues related to the causes of groundwater contamination
in their respective areas. It can be generally observed that level of education is a decisive
factor in showing concern for the environment.

Public health issues are universally of greater concern than the environment.
In general interviewees were wary of the effect of consuming contaminated water
because of their familiarity with health risks. Water-related illnesses are prevalent

in most developing countries, particularly in Sub-Saharan Africa. The general lack of concern over groundwater contamination among poorly educated focus group participants/survey respondents, was related to the potable status of their current water supplies. It also, however, relates to their increased concern of other socioeconomic issues, which affect their lives, in particular poverty. In this context, it is important to note that most participants and households surveyed live on less than the global benchmark of $1 per day, indicating extreme poverty. As previously noted the poor level of education plays a significant role in the ability of low income individuals to make informed decisions on issues related to groundwater contamination.

The common causes of groundwater contamination drawn from the interviews and focus groups are largely due to the widespread utilisation of pit latrines and open dumpsites, commercial activities, and agricultural practices. Domestic and commercial wastes are prevalent and widespread, while agricultural wastes are also generated, albeit in smaller amounts. The population density is estimated to be around 300–400 inhabitants per square kilometer, with a high number of inhabitants per household. The household survey revealed that 48.3% of the respondents affirm that pit latrine is the biggest causal factor of groundwater contamination, open dumpsites was next in rank with 28.5%. Other sources, such as domestic wastewater, tanneries, and dyeing works constitute about 15.3% and chemical and fertiliser application upstream of the residential areas make up the remaining 8%.

Open dumping and burning of all forms of waste in pits and in open spaces are common. The preference of these methods in the area has been practiced for a very long period. As previously identified it is obvious that the general public have little regard for the environment due to the predominant lack of awareness. Adequate waste collection facilities are missing and this has greatly influenced the attitude of the people towards poor waste disposal practices. Thus, it can be concluded that an attitude of indiscriminate waste disposal exists among the people.

The prevalence of these contamination sources in the study are is due to the cultural affiliation of the people towards on-site sanitation facilities, the unequal service provision rendered by the government, poverty, low level of public awareness, and lack of hygiene education among others. Thus, reversing these trends will require a shift from the current system to a more integrated and sustainable one.

4.5   SOLUTIONS

Solutions to the intractable issue of groundwater contamination in the study area requires an integrated strategy and are urgently required. Many other developing nations, particularly in Sub-Saharan Africa have similar issues. There are four overarching issues that need to be addressed:

4.5.1   Educating the citizenry on groundwater protection

The first step in achieving a groundwater protection system is educating the population to create an awareness among the population on the benefits of safe, clean water and the environment. The water sources needed for future development and population growth are being degraded by current waste disposal practices. Furthermore, the issue of pollution needs to be made aware of this to help curb contaminating practices. At the school level the State Government, through the Ministry of Education and the State Primary Education Board, has an important role to play by reviewing the current curriculum and ensuring teachers have the relevant training to boost environmental education in schools. Presently, the National School Curriculum only recognises health education and social studies at pre- and post-primary school levels. It is essential to incorporate environmental education into the current curriculum at all levels.

At the community level improved education on these issues is also required. Community associations, especially women and youth groups, farmers, and other relevant groups should be formed to work with the local community leaders in collaboration with the local authorities. In particular, NGOs, various State Ministries (Education, Environment, Water Resources, and Health) need to take a proactive role in advancing environmental advocacy and awareness creation among the population via community and interpersonal networks.

The general public must be adequately informed. The public cannot be expected to cooperate fully in relation to complex societal problems that are beyond their knowledge. The public needs to be informed of the risks associated with improper waste disposal and contaminated groundwater now and for the future.

4.5.2 Provision of adequate legislation and community rules

The current legislative framework is clearly not producing sustainable practices. Authorities must introduce legislation that will regulate groundwater development and constrain the activities that might compromise groundwater quantity and quality. This is because comprehensive water legislation offers considerable advantages that provide a legal basis for the effective and sustainable management of groundwater resources. The legislative framework needs to: (i) the clear explanation of the roles and responsibilities of all institutions and stakeholders active in the water sector; (ii) creation of water stakeholders and public-private partnerships; (iii) advancement of water provision mechanisms based on the social and economic value of water; and (iv) pollution control through the formulation and enforcement of communal rules that foster embracing solutions to the prevailing problems and individual initiatives at the community level.

The new legislation should provide for better gender-balanced involvement in private water supply regulation. Water supply legislation is out of date and controlled by males. It was identified in interviews that women are typically more responsive to environmental and social issues. Legislation should clearly state the terms for planning, allocation, and conservation of groundwater resources, as well as stakeholder interaction between local communities and institutions.

The State Legislative Assembly should ensure the provision of legislation that will regulate waste disposal and indiscriminate use of chemical fertilisers and pesticides on agricultural land in proximity to vulnerable groundwater resources. State environment protection agencies should be conferred with the constituted authority of enforcement and prosecution of perpetrators of inappropriate waste disposal practices and chemical applications. Likewise, the local community leaders, including traditional rulers, should be supported by the local authorities to persuade the local residents to support this scheme, and the responsibility of enforcing at a local community level.

Provisions should be put in place to require and resource the lining of all pit latrines in the existing and new development areas by the state (Ministry of Environment and the State Urban Development Board). The proliferation of pit latrines is not regulated, most of the pit latrines are not lined, so they will continuously contaminate groundwater resources of urban areas. Groundwater supplies in urban areas are import water supplies now and will become increasing important as the population grows as surface water resources become fully utilised. An integrated urban waste disposal (sewage) system is preferable, but is not possible in the current economic situation.

Presently, there are no fees and services associated with on-site household sanitation facilities and the state utilities are not involved in their control. In this regard, the relevant state institutions, in collaboration with the municipal councils, should recognise the need to be involved in regulating on-site household or community-based sanitation facilities. In addition, a long-term plan for the maintenance of on-site sanitation facilities should be supported by a budget. So, the authorities need to urgently assess the real requirements of the appropriate sanitation systems for each area and put plans, policies, and budgets in place to maintain these systems if they are to avoid additional groundwater contamination into the future.

In achieving progress, the state assembly should provide legislation that will prohibit construction of unlined pit latrines in residential areas. Legislation should also be put in place. That will compel households to pay sanitation fees (at least 1% proportionate to the cost of their development), which will be used to subsidise improved latrines with proper technical standards that suits the local conditions.

4.5.3 Waste management

Developing a robust waste management framework that considers the ethics, beliefs, and cultural norms of the people is essential. For this reason, the state and local governments and all other relevant institutions should adopt and implement programmes that will empower local women and youth groups through beneficial waste management activities. This has multiple benefits as it will ensure the protection of groundwater resources and the environment, which will help to prevent illnesses related to poor sanitary conditions. As an ancillary benefit it will create employment opportunities for jobless women and youths who are typically the lowest income earners.

In this regard, a potential site should be chosen as the first pilot scheme and a special task force committee, with state officials, representatives of the local residents and water users, group of youths and women, community leaders, and the local *Ulamas*, as well as a local NGO and professional groups, should be formed by the government to mainstream this policy.

4.5.4 Institutional integration and streamlining of responsibilities

The existing structure (top-down governance) is a major impediment and often results in inconsistency of government policy implementation. Therefore, a more integrated governance framework that brings together the relevant stakeholders (government ministries, water user groups, academia/technical experts, and all other relevant institutions) should be put in place, so that water and waste management are handled as a subsystem of a larger planning group, each impacting on the other. Additionally, the institutional framework for solid waste management must be addressed, with a view to bringing together the relevant institutional players and clarifying their responsibilities.

In achieving this, the current uncoordinated practices of water and waste management and urban planning as well as other municipal services must be integrated. This is because these presumed separate entities are interrelated implicitly and need to be managed holistically if environmental degradation and groundwater contamination are to be prevented. The piecemeal approach currently employed is not working.

For example, the Ministries of Water and the Environment should work with the university and any available technical experts to identify and map all of the potential pollution sources within cities, in order to protect and effectively manage the water resources of the area. The water resources ministry, in collaboration with the university, must carry out a comprehensive monitoring programme to identify trends in water quality and quantity. This information must be made accessible to the Ministry of Health and all relevant agencies and stakeholders. Likewise, the urban development board can work closely with the Ministry of the Environment in formulating policies that will mitigate the impact of new development on the environment. This coordination will contribute significantly and will lead to the development of sustainable pollution prevention and control strategies.

4.5.5 National and regional cooperation, and integration

To ensure sustainable management of groundwater resources in the Chad Basin and other areas across Sub-Saharan Africa, there is the need for greater ties between the various national and regional institutions. For example, regional disparities exists in cooperation among the riparian countries of the Lake Chad Basin (Nigeria, Niger, Chad, and Cameroun). These countries have signed a multitude of multilateral agreements for the sustainable management of transboundary aquifers. Similarly, the countries in West Africa (Economic Community of West African States) can cooperate with others in the eastern (East African Community) and southern (Southern African Development Community) Africa region for the development and adoption of the Africa Regional Action Plan on Integrated Water Resources Management (IWRM). Thus, local groundwater protection activities need to be planned in harmony with a broader regional policy framework.

4.5.6 Further commitment by external development partners

There is the need for greater ties and commitment between external groups (international development partners) and the federal government of Nigeria. There are number of international development partners working in Nigeria, providing support in health, education, rural development, social and development issues, and water-sanitation. However, it is imperative to increase the existing bilateral ties especially with the local community. Development partners need to deepen their commitments at the community level, especially in the areas of education, capacity building, and advocacy on sustainable groundwater management issues.

Also, development partners needs to further support the state and national institutions in the areas of manpower development, international and local policy development, and institutional capacity building among others. Increased institutional capacity at the federal, state, and local levels will ensure sustainable management of groundwater resources.

4.5.7 Additional commitment by the various tiers of government

The federal, states, and local governments have committed to improving the access to safe, clean, and affordable water in the country. However, despite their commitments the Millennium Development Goals for access to water and sanitation remain unrealistic to attain now and in the near future. Thus, the various governments need to fully comprehend the need for prioritising the water resource management agenda in their development policies. Also, the various governments must ensure effective implementation of water policies and strategies and the strict enforcement of water legislation. For the sake of sustainable water resource management it is equally important, to ensure that there are adequate returns from cost recovery to finance data collection, monitoring of system status, and resources management.

4.6 CONCLUSIONS

Stakeholder exclusion in the management of groundwater is a key feature of the current system. The major groundwater contamination problems are mainly attributed to the impact of pit latrines, open dump sites, and other non-point sources across the case study area. Most strategic stakeholders are familiar with the environmental problems, while majority of the primary stakeholders have limited knowledge about groundwater contamination issues. There is the need to move toward a more integrated and participatory groundwater management system that involves all the relevant stakeholders. Strategies such as, education and public awareness, enlightenment campaigns, institutional integration and streamlining of responsibilities, women and youth empowerment through waste management, and the provision of adequate legislation and communal rules will ensure the sustainability of the new system. These strategies should be vigorously pursued if change is to occur both in Nigeria and other developing nations.

ACKNOWLEDGEMENTS

The author wishes to thank the Petroleum Technology Development Fund for funding the study, Professors Chris Jefferies and Joseph Akunna of the Abertay University of Dundee, UK for their support, and all the stakeholders that take part in the study.

REFERENCES

Adelana, S.M. (2006). A quantitative estimation of groundwater recharge in parts of the Sokoto Basin, Nigeria. *Environmental Hydrology* 14(5), 105–119.
Adelana, S.M., MacDonald, A.M. (2008). Groundwater research issues in Africa. Applied Groundwater Studies in Africa. *IAH Selected Papers on Hydrogeology* 13, 43–64.

Africa Infrastructure Country Diagnostic (2011). *Africa's Infrastructure: A Time for Transformation. African Development Bank Group*, World Bank.

Akujieze, C.N., Coker, S.J., Oteze, G.E. (2003). Groundwater in Nigeria – a millennium experience-distribution, practise, problems and solutions. *Hydrogeology* 11(2), 259–274.

Bakari, A. (2014). Hydrochemical assessment of groundwater quality in the Chad Basin around Maiduguri, Nigeria. *Journal of Geology and Mining Research* 6 (1), 1–12.

Bakari, A., Akunna, J.C., Jefferies, C. (2014). Towards sustainable groundwater management in the south-western part of the Chad Basin, Nigeria: a stakeholder perspective. *Journal of Applied Science & Technology* 4 (25), 3727–3739.

Barber, W., Jones, D.G. (1960). The Geology and Hydrogeology of Maiduguri, Borno Province. *Records of the Geological Survey of Nigeria*, pp. 5–20.

British Geological Survey (2003). *Groundwater Quality Fact Sheet: Nigeria*. British Geological Survey, Keyworth.

Bunu, M.Z. (1999). Groundwater Management Perspectives for Borno and Yobe Sates, Nigeria. *Journal of Environmental Hydrology* 7 (19), 1–10.

Chilton, P.J. (1999). Groundwater recharge and pollution transport beneath waste water irrigation: the case of León, Mexico. 153–168 in Groundwater pollution, aquifer recharge and vulnerability. Robins, N. S. (editor). *Geological Society of London Special Publication, No. 130*, 105–121.

Eni, D.I., Obiefuna, J.N., Oko, C., Ekwole, J. (2011). Impact of urbanization on subsurface water quality in Calabar municipality, Nigeria. *International Journal of Humanities and Social Science* 110, 167–173.

Foster, S., Lawrence, A., Morris, B. (1998). Groundwater in Urban Development – Assessing Management Needs and Formulating Policy Strategies. *World Bank Technical Paper 390*. World Bank, Washington, DC, USA.

Furon, R. (1960). *Geology of Africa, second edition*. Payot, Paris.

Goldface–Irokalibe, I.J. (2002). Water Management in Federal and Federal–Type Countries: Nigerian Perspectives. *Africa Portal-a project of the Africa initiative.*

Hanidu, J.A. (1990). National growth, water resources and supply strategies in Nigeria in the1990's. *Water Resources J. Nigeria Association of Hydrogeologists*, 1, 1–6.

Hare, M., Pahl-Wostl, C. (2002). Processes of social learning in integrated resources management. *Journal of Community & Applied Social Psychology* 14(3), 193–206.

Isiorho, S.A., Matisoff, G. (1990). Groundwater recharge from Lake Chad. *Limnology and Oceanography* 35(4), 931–038.

Jacobsen, M., Webster, M., Vairavamoorthy, K. (2012). The Future of Water in African Cities: why waste water? *Direction in Development*, Washington, DC, World Bank.

Jaekel, D. (1984). Rainfall patterns and lake level variations at Lake Chad: in climatic changes on a yearly to millennial basis, in *Geological, Historical and Instrumental Records*, Morner, N. & Karlen, W., (Eds) D. Reidel Publishing Company, Dordrecht, pp. 191–200.

Lloyds, J.W. (1994). Groundwater management Problems in the developing World. *Applied Hydrogeology* 9(4), 35–48.

MacDonald, A.M., Bonsor, H.C., Calow, R.C., Taylor, R.G., Lapworth, D.J., Maurice, L.O., Dochartaigh, B.E. (2011). *Groundwater resilience to climate change in Africa, British Geological Survey Open Report* (p. 31). OR/11/031.

Maduabuchi, C., Faye, S., Maloszewski, P. (2006). Isotope evidence of palaeorecharge and palaeoclimate in the deep confined aquifers of the Chad Basin, NE Nigeria. *Environment* 370(1), 467–479.

Mcgrath, D., Zhang, C. (2003). Spatial distribution of soil organic carbon concentrations in grassland of Ireland. *Applied Geochemistry*, [online] 18, 1629–1639. Available from: doi: 10.1016/S0883-2927(03)00045-3 [Accessed 10th December 2014].

McNamara, C. (1999). *General Guidelines for Conducting Interviews*. [Online] 12 (11–13). Available from: http://www.mapnp.org/library/evaluatn/intrview.htm [Accessed 3rd May 2014].

Miller, R.E., Johnston, R.H., Olowu, J.A.I., Uzoma, J.U. (1968). Groundwater hydrology of the Chad Basin in Borno and Dikwa Emirates, with special emphasis on the flow life of the artesian system. *USGS Water Supply Paper* 1757.

Mitchell, R.K., Agle, B.R., Wood, D.J. (1997). Toward a theory of stakeholder identification and salience: Defining the principle of who and what really counts. *Academy of Management Review* 22(4), 853–886.

Morgan, D.L. (Ed.). (1997). *Successful focus groups: Advancing the state of the art* or ritualized research? *Family practice*, 19(3), 278–284.

Naik, P.K., Tambe, J.A., Dehury, B.N., Tiwari, A.N. (2008). Impact of urbanization on the groundwater regime in a fast growing city in central India. *Environmental monitoring and assessment* 146(13), 339–373.

National Population Commisions (2006). *Provisional 2006 Nigeria Census Results*. Abuja, Nigeria.

Nwankwoala, H.O. (2011). The role of communities in improved rural water supply systems in Nigeria: management module and its implications for vision 20: 2020. *Journal of Applied technology in Environmental Sanitation* 1(3), 295–302.

Obaje, N. (2009). *Geology and Mineral Resources of Nigeria*. Springer-Verlag, Berlin and Heildelberg. ISBN 978-3-540-92684-9.

Odada, E.O., Oyebande, L., Oguntola, J.A. (2006). Lake Chad. Experience and Lessons Learned Brief. *Lake Basin Management Initiative (LBMI) Experience and Lessons Learned. LCBC working paper 102.*

Offodile, M.E. (1992). *An approach to groundwater study and development in Nigeria. Mecon Services Ltd.* pp. 66–78.

Oteze, G.E., Fayose, S.A. (1988). Regional development in the Hydrology of Chad basin. *Water Resources* 1(1), 9–29.

Patten, M.Q. (2001). *Qualitative evaluation and research methods (2nd ed.)*. Newbury Park, CA: Sage.

Putra, D., Baier, K. (2008). Impact of Urbanization on Groundwater Recharge – The Example of the Indonesian Million City Yogyakarta, In: UN Habitat- United Nations Settlement Programs: Fourth session of the World Urban Forum, Nanjing, China, Documentations of Germany's Contribution to a Sustainable Urban Future.

Rijswlk, K. (1981). Small community water supplies. *IRC Technical Paper Series*, No. 18. The Netherlands.

Rotmans, J., Kemp, R., van Asselt, M. (2000). *Transitions and transition management, the case of an emission-free energy supply*. International Centre for Integrative Studies, Paper Series, No. 10. The Netherlands.

Schnieder, J.L., and Wolff, J.P. (1992). Geological map and hydrogeological maps at 1: 1,500,000 of the Republic of Chad – Explanatory Memorandum. *BRGM Document No. 209*. BRGM, Orleans.

United Nations (1988). Ground water in North and West Africa. *Natural Resources/Water Series No. 18*. The United Nations, New York; 198.

Hausa, V.O.A. (2014). Map of Nigeria showing Maiduguri. [Online] 10 (1–3). Available from: http://www.voahausa.com/a/1646766/i6.html [Accessed 3rd December 2014].

Wakida, F.T., Lerner, D.N. (2005). Non-agricultural sources of groundwater nitrate: a review and case study. *Water Research* 39(1), 3–16.

World Health Organisation; UNICEF (2000). *Global Water Supply and Sanitation Assessment 2000 Report*. Geneva and New York, World Health Organisation and United Nations Children's Fund.

Chapter 5

Governance and management of hydrogeological impacts of unconventional hydrocarbons in Australia

L. Lennon[1] & W.R. Evans[2]

[1]Jacobs (formerly SKM), Melbourne, Victoria, Australia
[2]Formerly SKM, now Salient Solutions, Canberra, ACT, Australia

ABSTRACT

The expansion of the onshore gas industry (shale, tight, and coal seam gas) in recent years has been almost exponential. This has attracted significant attention from both policy makers and the community, generally driven by the potential for impacts on other industries reliant on groundwater resources and the broader environment. The key hydrogeological risks associated with onshore gas development relate to well integrity issues, hydraulic fracturing, and water management. Risks associated with well integrity issues and hydraulic fracturing are governed by the relevant oil and gas regulations and are also reliant on implementation of good practices. However, risks associated with hydrogeological impacts of gas and water extraction cross into the water resource planning sphere where a range of different approaches can be applied.

Fundamental to water resource planning is the need to manage potential third party impacts for both surface water and groundwater users over time. This chapter suggests that management of third party impacts in the context of the unconventional hydrocarbon industry needs to be considered in the context of the level of connectivity with other resources. In systems that are highly connected it is critical to incorporate water use and potential impacts associated with gas development into the management planning and licensing processes. This is likely to be the best way to manage third party impacts and ensure groundwater use and impacts are within sustainable limits.

However, there are a number of challenges unique to onshore gas development including the extent of time lags for impacts to occur, the large number of wells required, the variability in extraction rates, and the sometimes large quantities of water required for unconventional hydrocarbon projects in the context of available water. There are technical issues to solve around how groundwater deposits were (or were not) accounted for under the defined sustainable yield or total consumptive pool for the resource in many jurisdictions.

5.1 INTRODUCTION

Oil and gas production has traditionally been regulated and licensed via a Petroleum and Gas Act (or similar) enacted within each State. Historically, conventional oil and gas production were predominantly offshore activities or were confined to remote sedimentary basins with little competition for water resources. In recent years, the development of unconventional oil and gas, such as shale, tight, and coal seam gas (CSG), has become technically feasible and economically viable. Some of Australia's unconventional reserves are located close to densely populated areas or agricultural/ pastoral activities and have the potential to impact groundwater resources that are already in a mature state of use and the broader environment.

This chapter discusses issues associated with managing and regulating hydrogeological impacts of unconventional gas development and outlines the governance arrangements around Australia. Section 5.2 provides background with a brief overview about unconventional gas development in Australia. The key hydrogeological risks associated with unconventional gas are presented in Section 5.3 and Section 5.4 outlines the current regulatory regime that governs and manages these risks. Section 5.5 provides a discussion on the merits and difficulties of a range of potential management strategies that could be adopted. Conclusions are presented in Section 5.6. This paper covers the impacts on groundwater quality and quantity associated with unconventional gas development, including potential impacts from well integrity, hydraulic fracturing, and water management. A summary of Australian regulations of this critical, emerging hydrogeological issue is useful for not only Australians but for an international audience. Many jurisdictions are still grappling with the best way to regulate the unconventional hydrocarbon industry and this paper summarises not only issues, but proposes possible solutions.

5.2 UNCONVENTIONAL GAS IN AUSTRALIA

There are two primary sources of natural gas: conventional and unconventional. Conventional gas refers to gas trapped in multiple, relatively small porous zones in various rock formations, like sandstone. Conventional gas exists as free gas, which has migrated away from its source rock and is trapped in a reservoir by an impermeable layer or seal. It is typically significantly easier and more cost effective to extract. In comparison, unconventional gas remains in situ in the formation in which it was produced and is held there by pressure and/or a lack of permeability. Technological advances in horizontal drilling and hydraulic fracturing have made unconventional gas supplies more commercially viable (SKM, 2013).

According to Geoscience Australia (2015) there are three main types of unconventional gas:

- **Coal seam gas (CSG)**, also known as coalbed methane (CBM), is natural gas found in coal seams. The gas is adsorbed to the coal matrix and held there by hydrostatic pressure. Reducing the pressure in the coal seam by dewatering releases or desorbs the gas from the coal matrix. As pressure is reduced, gas flow increases and water flow rates decrease over a period of a few months depending on the hydrogeological conditions. CSG is the shallowest unconventional gas, typically found between 300 m to 1000 m depth.

- **Tight gas** is more similar to conventional gas in that it has migrated away from the source rock and is found in sandstone, sands, and carbonate deposits that have a very low permeability. Tight gas reservoirs are generally deeper than CSG at depths typically ranging between 1000 and 3000 m. The gas is extracted from the formation, which has a low permeability and is required to be hydraulically fractured to increase the permeability to release the gas.

- **Shale gas** is typically found at even deeper depths (2500 to 4000 m) in the fine-grained sedimentary rock called shale. Shale gas is more comparable to CSG, in that the gas remains within the source rock and has not migrated into a porous or more permeable reservoir.

Unconventional gas is typically more difficult and costly to extract for various reasons depending on the resource and geological setting. In the case of CSG, groundwater in varying quantities is also produced (i.e. coproduced water) and requires management and disposal. Whereas for tight and shale gas, hydraulic fracturing is required to increase the permeability of the host rock, which often means tight and shale gas developments are net water users. The higher production costs associated with unconventional gas means it is generally only economically feasible with higher gas prices. When gas prices fall, industry tends to focus on the higher permeability reservoirs that require less effort to produce gas.

Unconventional gas is not the same as unconventional oil. Examples of unconventional oil reservoirs include oil shales, oil sands, extraheavy oil, gas-to-liquids, and coal-to-liquids (Geoscience Australia, 2015). The following sections outline in more detail unconventional gas exploration and development in Australia.

5.2.1   Coal seam gas

In Australia, exploration for CSG commenced in the Bowen Basin, Queensland, in 1976 and commercial production began in the same area in 1996. CSG is now an integral part of the gas industry in eastern Australia. Most of the CSG reserves in Australia are located in New South Wales and Queensland and together these make up 78% of the Eastern Gas Market reserves (Geoscience Australia, 2010).

Queensland's CSG reserves are located in the Surat and Bowen Basins. A major driver for the growth of CSG in Queensland was the State Government's commitment to have 13% of the State's power grid to be generated by gas by 2005 (Geoscience Australia, 2015). This requirement has subsequently been increased to 15% by 2010 and 18% by 2020. The increased rate of gas production led to a number of proposed liquefied natural gas (LNG) export terminals, further increasing production as a function of international demand. The basin with the greatest potential for CSG development and currently the subject of major development is the Surat Basin in south central Queensland.

In New South Wales, CSG investigation and production is relatively small in scale compared to current operations in Queensland. AGL are currently producing CSG in the Sydney Basin and exploration is underway in the Gloucester Basin. Exploration is also underway in the Gunnadah Basin (Santos) and in the Clarence–Moreton Basin (Metgasco).

In Victoria, CSG exploration is in its infancy as a result of moratoriums on exploration and hydraulic fracturing established in 2013 by the State Government, are currently still in place. The brown coal deposits in the Gippsland Basin are the most likely prospective for CSG. However, it is yet to be proven if there is an economical resource. Some coal seam gas exploration is being undertaken in South Australia in the Cooper Basin and the Eromanga Basin (DMITRE, 2012). There is limited exploration for CSG in other states and territories.

5.2.2   Tight gas

Tight gas is similar to conventional gas in that it refers to gas that has migrated away from its source rock. The difference between tight and conventional gas relates to the permeability of the host rock. Conventional gas is trapped in a high permeability reservoir and is held there by a 'seal' or low permeability layer. Tight gas, on the other hand, is trapped in a low permeability formation and requires hydraulic fracturing or stimulation to increase the permeability to facilitate gas production. It is somewhat subjective about what constitutes tight gas. Permeabilities of individual tight gas reservoirs are highly variable, depending on the region and can range between 0.0001 and 0.1 millidarcy (DMP, 2013).

Given the variability in permeability, tight gas reservoirs around Australia are less well defined and exploration has been limited. Two companies have drilled onshore tight gas wells in the Northern Territory and Victoria. Central Petroleum drilled a well into the tight gas sands plays in the Pedirka Basin in Northern Territory and Lakes Oil has drilled several wells in the Gippsland Basin, Victoria. Over the last decade, Lakes Oil has explored the tight gas reserves in the Strzelecki Formation below the Latrobe Group. Exploration for tight gas in Victoria in recent years has been constrained by the moratoriums currently in place.

5.2.3   Shale gas

The shale gas industry in the United States has grown substantially over the last decade or so as a result of technological advancements, such as horizontal drilling and hydraulic fracturing, which have also facilitated tight gas production. While these technologies have become more accessible in Australia in recent years, shale gas exploration and production, although the former widespread, the industry is in its infancy in Australia. Exploration to date indicates that shale gas resources could exist in many sedimentary basins in Western Australia, Northern Territory, South Australia, Queensland, and Victoria (refer Figure 5.1).

The most well-known and arguably the most prospective basin for shale gas is the Cooper Basin in central Australia, spanning the South Australian and Queensland borders. The basin is Australia's most mature onshore basin. Conventional gas development has been occurring for over 40 years and the region is also attractive due to existing infrastructure, which currently supplies gas to South Australia, New South Wales, Queensland, and Victoria (CSIRO, 2012). There are several companies exploring for unconventional hydrocarbons in the Cooper Basin, although Santos has the only gas producing shale gas well (Yeo, 2012). Surrounding the Cooper Basin are a series of Paleozoic basins with shale gas potential in South Australia, Queensland and Northern Territory (Frogtech, 2013). Other prospective shale gas resources in South Australia include the Arckaringa, Otway, Pedirka, Simpson, and Warburton Basins (DMITRE, 2012).

In the Northern Territory two basins have prospective shale gas, the McArthur Basin and the Georgina Basin. The McArthur Basin is located in northern Northern Territory and there a number of deeper sub-basins within the basin with the most important being the Beetaloo Sub-basin (Frogtech, 2013). Limited exploration to date and existing infrastructure could limit exploration and development in the Beetaloo Sub-basin (CSIRO, 2012). The Georgina Basin is a region of proven oil

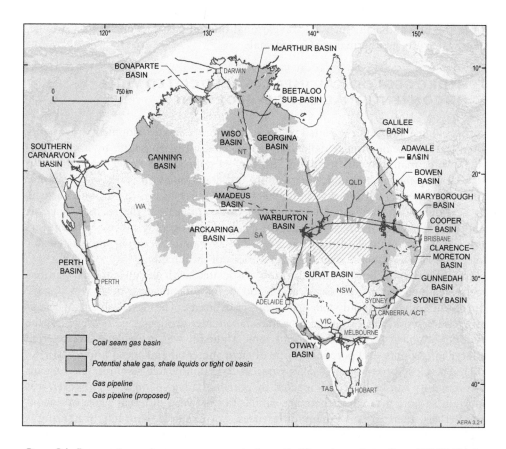

Figure 5.1 Prospective onshore gas resources in Australia (Geoscience Australia and BREE, 2014).

and unconventional gas exploration has only recently commenced. The Norwegian firm Statois is currently exploring for shale gas with 10 to 20 wells proposed to be drilled by 2017 (Yeo, 2012).

Exploration for shale gas has commenced in the Canning Basin in Western Australia, targeting the Goldwyer Shale. However, CSIRO (2012) highlighted that the region is remote with a low population, little or no industrial activity, and a limited road network, which means any gas production would require significant investment. In comparison, the Perth Basin has existing conventional oil and gas fields and given the close proximity to the Perth market, existing infrastructure and dwindling conventional reserves, means unconventional gas is becoming attractive.

In Queensland, in addition to the Cooper Basin, the Galilee Basin has long been recognised as a potential resource and exploration for shale gas is targeting the Toolebuc Shale. CSIRO (2012) noted the key challenge in the Galilee Basin will be the significant investment required in infrastructure. Exploration activities are also underway in the Eromanga and Maryborough Basins.

The Otway Basin in Victoria has been a conventional oil and gas resource since the 1950s and there are a number of potential shale gas resources (Frogtech, 2013).

Frogtech (2013) highted the multiple users and conflicting demands that currently exist in the Otway Basin, such as conventional oil and gas, carbon capture storage, groundwater use, geothermal, CSG, and shale.

5.3 HYDROGEOLOGICAL IMPACTS ASSOCIATED WITH UNCONVENTIONAL GAS DEVELOPMENT

Hydrogeological impacts associated with unconventional gas development are broadly related to: large scale aquifer depressurisation, well integrity issues, hydraulic fracturing, water management (including disposal of coproduced water via reinjection and water supply for hydraulic fracturing), and contamination issues. This chapter defines hydrogeological impacts as impacts that occur beneath the ground surface. It is acknowledged that some activities can cause impacts at the surface, such as chemical spills, which indirectly impact the shallow hydrogeological environment (i.e. water table), and induced seismicity as a result of hydraulic fracturing and wastewater reinjection. However, these issues are not a focus of this chaper.

Community concerns around impacts associated with unconventional gas has largely stemmed from CSG exploration and development and issues, which have occurred in some tight and shale gas projects overseas. The development of a regulatory framework was perceived by some stakeholder groups to be slower than the actual development of the industry and some argued it was inadequate, considering the large uncertainty and potential scope of environmental and social impacts (EDO, 2011). The close proximity of CSG resources to groundwater resources also means the community can be directly impacted by potential adverse impacts. However, it is important to note that while there are similarities between different types of unconventional gas, there are also many differences. For example, hydraulic fracturing is not always required for CSG and large volumes of groundwater (coproduced water) are often produced and require management (i.e. treatment and disposal). In some cases this coproduced water is reinjected into aquifers if suitable hydrogeological conditions exist and it is feasible to do so under State regulation. Shale and tight gas, on the other hand, requires water for hydraulic fracturing and produces very little water during gas production. This means they are net water users and locating a suitable water resource for use during fracturing operations can be a limiting factor in remote parts of Australia. The discussion below outlines potential hydrogeological impacts associated with unconventional gas development.

5.3.1 Well integrity

Well integrity is a term used to describe the standards required for construction of wells so that during the whole life cycle of the well movement of uncontrolled movement of fluids, solids and gases into the broader subsurface or surface environment is prevented. Well integrity is fundamental to protect the aquifers that the well intersects for the full life cycle of the well. Exploration and production of unconventional gas may involve drilling thousands of wells that insect shallower groundwater resources. If well integrity is not maintained, or wells are not abandoned properly, there is potential to cause significant impacts to regional groundwater resources. This has the potential to affect existing and future users (including the environment) on an ongoing basis. The key to maintaining well integrity is the reliability of the cement or other seal around the well casing to effectively isolate the targeted zone from other hydrogeological layers. This applies to both well construction and decommissioning/abandonment.

The IESC (2014) published a background review on bore integrity in the water, mining, and coal seam gas industries. This review highlighted that in Norway the Petroleum Safety Authority found that the integrity of 18% of wells had either failed or had issues with their integrity (Birgit and Aadnoy, 2008). Davies *et al.* (2014) also conducted a review of well integrity issues in oil and gas wells and this outlines many statistics about failed wells. For example, the Pennsylvania state database has records for 3555 wells and their well barrier or integrity failure rate is 6.3%. Integrity issues ranged from failure to case and cement properly to excessive casing pressure. In offshore Norway, of the 406 wells examined by Vignes and Aadnoy (2010) 18% had well integrity issues (75 wells) and 11% of the failures were related to issues with the cement. In the UK, the Environment Agency has had nine reports of pollution from onshore gas wells between 2000 and 2013 and two of these were related to well integrity issues, specifically the cement around the well head (Davies *et al.*, 2014). There is no such information available in the public domain on similar studies being undertaken in Australia. However, the IESC (2014) noted a single case reported in the media in August 2012, where a coal mining exploration hole that had not been decommissioned/abandoned properly, started leaking gas which, subsequently started a small bush fire.

In Norway, key factors for well failure related to operational decisions made during abnormal situations, poor well design, and the inability to account for rare events that can lead to major incidents (Birgit & Aadnoy, 2008). It must also be noted that many aquifers have water chemistry, which can be aggressive towards materials used in well construction, particularly cement. Even FRP (fiberglass) and stainless steel casing will degrade over time. Seismic activity, both natural and induced through gas production activities, could also cause well integrity failures. There is significant potential for wells to lose integrity over long (+20 year) time frames, but these legacy issues have not been a focus of significant research or regulation.

The American Petroleum Institute (API, 2009) states that although the selection of materials for cementing and casing is important, it is secondary to cement placement. The key to good cementing is good operational practices (Nygaard, 2010; Corneliussen *et al.*, 2007; Bourgoyne *et al.*, 1999). Cement for petroleum and gas wells are engineered products that are governed by the American Petroleum Institute technical standards (IESC, 2014). The petroleum industry uses Portland cement with several additives, such as density reduction materials, viscosifiers, accelerators, and retarders to refine the cement slurry (Nygaard, 2010). The durability of cement is obviously variable, depending on the design, construction works, additives, and exposure conditions. However, Naik & Kumar (2003) discuss options to design structures with a lifespan of 1000 years. There is very little information on the design lifespan of well construction and abandonment in the public domain. Much of this information is held by the industry. Design lives for well construction are not specified in regulations and guidelines as these are focussed on specification of correct materials and procedures.

One of the key findings from a workshop on well integrity for the long term geological storage of CO_2 held in Texas was that it is not possible to promise a leak-free well (Pearce, 2005). Pearce highlighted that the focus should be on industry constructing the best wells possible, rather that presenting well designs and constructions as providing a leak-proof solution. Pearce suggested that it may not be necessary to demonstrate well integrity for 1000 years and instead provide shorter term integrity (e.g. 100 years). If shorter terms can be proven, it could be extrapolated over longer time frame.

5.3.2 Water management

Water management issues differ depending on the type of unconventional gas development. CSG poses the greatest risks to water resources, given its typically close proximity to fresh (generally shallow) groundwater resources. For CSG development SCER (2013) broadly categorise the issues as aquifer depletion and contamination of water resources. Both can affect existing groundwater users, groundwater-surface water interactions, and groundwater dependent ecosystems. CSG is held in the coal seam by hydrostatic pressure and does not require a low permeability 'seal' to hold the gas in place like other forms of gas. Gas development from coal seams requires large scale depressurisation to release the gas adsorbed to the coal matrix and depending on the permeability of the coals, large volumes of coproduced water can be generated. Depressurisation of regional groundwater systems will occur where large volumes of groundwater are required to be extracted, particularly where good hydraulic connectivity between the seams and the adjacent aquifers exists. The impacts of depressurisation in the coal seams will vary from site to site depending on the degree of connection between aquifers and the hydraulic properties of the coal seam. In comparison, large-scale aquifer depletion is not required for shale and tight gas as these formations have low permeabilities and less water is produced during gas development. Shale and tight gas development is actually a net water user as water is required for hydraulic fracturing operations. Small-scale aquifer depressurisation may result if a groundwater supply is used for hydraulic fracturing operations and this may have direct impacts on groundwater–surface water interactions and groundwater dependent ecosystems. This is not, however, a direct result of the hydraulic fracturing process per se.

In addition to impacts on reduced groundwater levels where the coal seams are situated within a larger regional aquifer, such as the Surat Basin, the large-scale depressurisation of the coal seams has the potential to release gas into water boreholes that also intersect the coal seams. In the case of the Surat Basin there are minor coal seams embedded through most of the formations and many water bores intersect these coal seams. When the target coal seam is depressurised, the surrounding aquifers will also be depressurised to some extent and gas can be released (desorbed) from these minor coal seams and be detected in privately owned water bores (SKM, 2012).

There are also potential issues associated with the management or disposal of the large volumes of water generated during coal seam gas operations. There are a range of management options, including reinjection into the same or different aquifers, discharge to rivers or for beneficial use, such as agricultural use. The volume of coproduced water varies spatially and temporally. This can pose difficulties to some beneficial uses that require security of supply over longer timeframes (i.e. farming enterprises). Some of these management options require water treatment to improve the water quality. Treatment processes can produce brine, which either needs to be disposed of in an appropriate waste facility or in some cases reinjected into a deeper, typically already saline aquifer. As this paper is focussed on hydrogeological impacts, the water management options with the greatest risk to groundwater are reinjection of either coproduced water or brine.

For shale gas development, Cook *et al.* (2013) note that the key considerations regarding water resources relate to water extraction for water supply purposes and water discharge, as well as contamination of surface water or groundwater in association with these activities. They also highlight that because shale gas development is in its infancy in Australia, the volume of water required for hydraulic fracturing is not well understood. Typically, hydraulic fracturing in shale or tight gas formations will require an order of magnitude more water than CSG due to the greater depths and different hydraulic properties. However, the volume of produced water is significantly less from shale, which means less storage, treatment, and reuse options are needed for shale gas development.

Shale resources in the Cooper and Galilee Basins and CSG resources in Queensland exist in the Great Artesian Basin. The Great Artesian Basin is a confined aquifer system that is mostly artesian and supports many groundwater users and groundwater dependent ecosystems. In the past groundwater bores have been drilled into the Great Artesian Basin and have been allowed to flow uncontrolled. It was recognised in the early 1990s that groundwater pressures and levels were declining as a result of these free-flowing bores. As a result the Great Artesian Basin Sustainability Initiative (GABSI) was established to control the flows from these bores by capping them and replacing the irrigation infrastructure (DoE website, 2015). Gas development will need consider the protection of Great Artesian Basin aquifers from contamination and influence on the cumulative impacts of water requirements from the Great Artesian Basin.

5.3.3 Hydraulic fracturing

Hydraulic fracturing is a technique that has been employed in the petroleum industry globally for over 60 years and in Australia for over 40 years. It is the process by which hydrocarbon-bearing formations are 'stimulated' to enhance the flow of hydrocarbons to the well head and involves the injection of fluid (and other materials) under high pressure into a geological formation from, which hydrocarbons (oil and gas) are intended to be extracted. This stimulation creates or enhances permeability, including existing fluid and gas pathways. Typically, the process creates additional fractures in the reservoir rock and these can be held open for a period of time through the use of proppant (sand or a man-made equivalent). Hydraulic fracturing fluids are primarily composed of water (typically 90%), proppant (typically 9%), and chemical additives. The chemical additives used in the hydraulic fracturing fluid vary, depending on the application, the nature of the target gas formation, the individual developer, and in some cases legislation and regulations applicable to the region (SKM, 2012).

Hydraulic fracturing is required for most, though not all types, of unconventional gas development. Fracturing is needed for shale gas and tight gas development due to the low permeability in the target gas formation. In the case of CSG some, but not all gas fields have a high natural permeability due to the fractures in target formation cleats, which means that hydraulic fracturing is only required on occasion. In Australia, since 2000 8% of CSG wells have been hydraulically fractured. The industry estimated that between 25 and 40% of wells yet to be drilled for current CSG developments across Australia (mainly Queensland) may need some method of flow enhancement, including hydraulic fracturing (GISERA, 2011). In the Cooper Basin, around 70 wells have been hydraulically fractured over the period of development (SKM, 2013 cited in Cook *et al.*, 2013).

The US EPA (2011) identified four mechanisms by which hydraulic fracturing can cause or increase the potential for groundwater contamination, including:

- Failure of wells during the hydraulic fracturing process, which may create pathways by which contaminants can affect groundwater assets.
- Leakage of hydraulic fracturing fluids beyond the fracture zones from the target zone to adjacent formations.
- Mobilisation and migration of naturally occurring contaminants from the target zone to adjacent formations via fractures.
- Leakage of gas from target formations.

SCER (2013) broadly characterises key risks of hydraulic fracturing, which include excessive fracture propagation, resulting in potential groundwater contamination via fracture fluid leakage and increased connectivity between naturally occurring contaminants and groundwater resources.

Hydraulic fracturing operations also have the potential to cause groundwater and surface water contamination via a range of other mechanisms, including storage of hydraulic fracturing water at the surface, chemicals spills, and pipeline leakages and failures. King (2012) identified 20 key risks associated with hydraulic fracturing, including on-site spills and well integrity issues induced by hydraulic fracturing. Well failure can occur due to incorrect construction, poor seal construction in the annulus, high pressures, or corrosion. Proper construction of a well, correct use of materials as well as appropriate fracturing processes and techniques are all crucial to protect groundwater resources. Given the focus of this paper, issues relating to water storage, contaminant spills, and failure of surface infrastructure are not considered further.

A typical hydraulic fracturing injection event will range between 20 minutes and 4 hours, depending on the design (Cook *et al.*, 2013). After the hydraulic fracturing process has been completed production begins and both water and gas will flow. During this time around 15 to 50% of the hydraulic fracturing fluid is recovered (Cook *et al.*, 2013). The flow back water is either reused for subsequent hydraulic fracturing events or disposed of in accordance with regional regulations. Cook *et al.* (2013) highlights that particularly for shale and tight gas the hydraulic fracturing fluid that remains in the formation poses little or no environmental concern as it is trapped at great depth and typically cannot migrate rapidly from the formation. The different approaches to managing hydrogeological risks relating unconventional gas exploration and development in each jurisdiction are outlined below.

5.4 MANAGEMENT APPROACHES AROUND AUSTRALIA

Unconventional gas is a type of hydrocarbon or petroleum that is generally regulated in the same way as conventional oil and gas. In Australia the regulatory framework for the unconventional industry is complex, with a mixture of Commonwealth and State planning, environment, mining, water, and OH&S legislation applicable to gas developments dependent upon the characteristics of individual projects. The rapid development of the CSG industry in Queensland and New South Wales, coupled with increased publicity on the perceived and potential social and environmental impacts, prompted State and Commonwealth governments to initiate significant regulatory review. Since 2010, the Commonwealth and States have invested heavily in unconventional (CSG) gas regulation. Prior to this review and investment, the development of a regulatory framework was perceived by some stakeholder groups to be slower than the actual development of the industry. Some have argued it was inadequate, considering the large uncertainty and potential scope of environmental and social impacts (EDO, 2011).

Comprehensive regulations are now in place for conventional and unconventional (primarily CSG) gas operations. Cook *et al.* (2013) suggest that many features of the existing regulatory regimes are transferable to shale (and tight) gas development and related activities. Regulatory regimes and approaches to managing hydrogeological impacts are different in each jurisdiction and are discussed below.

5.4.1 Commonwealth

The Commonwealth Government has powers to review unconventional gas development (CSG specifically) approvals via the *Environment Protection and Biodiversity Conservation Act 1999* (EPBC Act) if the development will have a significant impact on a Matters of National Environmental Significance. Until recently the EPBC Act applied only where CSG activities were assessed as having a potentially significant impact on Matter of National Environmental Significance listed under the Act. However, the National Partnership Agreement on Coal Seam Gas and Large Coal Mining Development agreed to establish the Independent Expert Scientific Committee on CSG and Coal Mining (IESC) under the EPBC Act. The Commonwealth and signatory jurisdictions refer CSG projects that are likely to have a significant impact on Matters of National Environmental Signifiane, which now includes water resources to the IESC for advice. The National Partnership Agreement expired in June 2014, although it is intended that all signatories will continue to ensure the objectives of the National Partnership Agreement are met.

The Standing Council on Energy and Resources (SCER) endorsed the National Harmonised Regulatory Framework for Natural Gas from Coal Seams (SCER, 2013), which presents 18 leading practice strategies to manage impacts for CSG exploration and development. The leading practice strategies are designed to manage the potential risks associated with well integrity, water management, hydraulic fracturing, and chemical use in CSG exploration and development. As mentioned previously, many of these are relevant to shale and tight gas development. The framework outlining leading practice regulation is not intended to be a static concept (SCER, 2013). Effective implementation and intergovernmental cooperation will ensure the framework evolves as required. Jurisdictions are required to report back to SCER on progress in implementing the framework and on areas where existing legislation remains inconsistent with it (SCER, 2013).

5.4.2 Queensland

In Queensland, unconventional (CSG) gas extraction is primarily regulated through the *Petroleum & Gas (Production and Safety) Act 2004* or the *Petroleum Act 1923*, which are administered by the Department of Natural Resources and Mines (DNRM) via the granting of petroleum tenure. The *Petroleum and Gas (Production and Safety) Regulation 2004* sets out mandatory and recommended codes of practice. The Code of Practice for Constructing and Abandoning Coal Seam Gas Wells in Queensland is the primary means of preventing hydrogeological risks associated with well integrity issues (DNRM, 2013). Other unconventional gas wells are required to be constructed in accordance with the relevant state petroleum legislation.

Hydraulic fracturing risks are managed through the *Environment Protection Act 1994* via environmental management and water management plans (Hoare & Finn, 2014). A risk assessment is required as part of the environmental management plan and risks associated with hydraulic fracturing operations are included.

A licence under the Petroleum and Gas act also includes the right to take groundwater. The Queensland *Water Act 2000* was amended in December 2010 to introduce a new regulatory framework to manage the cumulative impact on water supply bores and springs from the extraction of groundwater by petroleum tenure holders, including the CSG industry. The adaptive management framework seeks to manage the cumulative impacts, resulting from the extraction of CSG water through the designation of Cumulative Management Areas (CMA) and requirements for the preparation of an Underground Water Impact Report (SKM, 2012). The framework makes provisions to include baseline monitoring and assessments, 'make good' agreements with land owners, dispute resolution process, and established the Office of Groundwater Impact Assessment (formerly the Queensland Water Commission) to manage the cumulative impacts of CSG activities. The Underground Water Impact Report framework also establishes responsibilities for petroleum tenure holders to monitor and manage the impacts caused by the exercise of their water rights, including a responsibility to make good impairments of private bore water supplies and protect springs via the Spring Impact Strategy (QWC, 2012).

In terms of the Underground Water Impact Report process, Queensland is focusing on the Surat and southern Bowen Basins, which have both been declared being part of the Surat CMA. The Underground Water Impact Report defines predicted impacts based on water level drawdown trigger thresholds and specific 'make good' arrangements will be required with consumptive users within these affected areas. Individual bore trigger levels (impact threshold) are defined as the amount of decline in water level in an aquifer or bore impacted by CSG operations, which could pose a risk to water supply from the bore as prescribed by law. The process for determining the level of acceptable risk inherent in the prescription of a trigger threshold is not outlined, which highlights a bigger issue around how acceptable impacts are derived. Water take associated with CSG development is currently being managed explicitly in Queensland and News South Wales. Other jurisdictions are aware of the issue but to date had no reason to specifically apply regulations to CSG operations.

5.4.3   New South Wales

In New South Wales, the activities associated with unconventional (CSG) exploration and development must take place in accordance with a petroleum title issued under the New South Wales *Petroleum (Onshore) Act 1991*. Petroleum titles contain standard conditions approved by Office of Coal Seam Gas, which was established within the

Department of Trade and Investment in February 2013 (Hoare & Finn, 2014). All regulation is specific to CSG in New South Wales. Two codes of practice were released in 2012, which govern all risks associated with well construction and abandonment and hydraulic fracturing activities (New South Wales Trade and Investment, 2012a,b). The code of practice relating to hydraulic fracturing or fracture stimulation covers the use of chemicals and sourcing of water for fracturing fluids and the protection of groundwater resources. Environmental risks associated with all CSG operations, including drilling and hydraulic fracturing, are assessed and managed under the *Environmental Planning and Assessment Act 1979*.

Impacts relating to groundwater take are regulated under the *Water Management Act 2000* where a water sharing plan is in place and where water sharing plans do not exist under the *Water Act 1912*. Licences for water take and disposal are granted under the relevant water act. Since March 2013, a water access licence is required for all petroleum activities (exploration and production) for both current and historical activities (albeit some exceptions) (Hoare & Finn, 2014). This approach is complicated by the two water acts that control the planning process depending on location. CSG development in New South Wales will typically be in areas subject to the *Water Act 1912* (rather than the *Water Management Act 2000*), hence these areas do not have robust long-term annual extraction limits due to the limited information available and lack of water sharing plans.

To further manage the potential impacts of CSG operations, the New South Wales Government implemented the Aquifer Interference Policy (Department of Primary Industries, 2012) under the New South Wales *Water Management Act 2000*, which aims to address the assessment and protection requirements for groundwater across the State. Under the Policy, any activity that can cause interference to an aquifer's values will require assessment and licensing. CSG production is such an activity. The Policy also requires that all mining and petroleum exploration activities (including CSG) to hold a water access licence where the volume of groundwater taken during these activities is greater than 3 ML/year. Where a number of different aquifers (or indeed water sources) are impacted, a separate licence is required for each aquifer/water source. The proponent must hold a licence, both for the duration of the extraction and the duration of impact after extraction has ceased. The Policy specifies the information that the proponent must provide, including predictions of the amount of water taken from each system, the approach to obtaining licences, managing low allocation years, and the approach to managing legacy issues. Where the relevant plan has a provision for unassigned water, then an entitlement can be granted, either by application or by tender/auction. Where no unassigned water is available under the relevant plan, then the proponent can purchase an entitlement via the water market, which is subject to the respective Water Sharing Plan. However, in many cases the total consumptive pool determination for each Plan area is not always robust and interactions between the resources in adjacent plan areas are not assessed. The Aquifer Interference Policy also sets out Minimal Impact Considerations for both highly and lesser productive aquifers. These essentially provide a threshold (similar to the Queensland bore trigger thresholds) below which there is deemed to be no diminution to a third party's rights to water.

As part of the approval process, the CSG activity also needs to pass the gateway test prior to the consideration of a development application. This gateway test, in part, considers issues related to protection of groundwater. Once past this decision, the proposed activity will need to be subject to an environmental assessment in the form specified in the *Environmental Planning and Assessment Act 1979* (New South Wales). There are three approval streams through this latter act, depending on the type of project development being considered. In all cases, issues related to impacts on water resources are included in the decision process. In addition, a project can be designated as being of State Significance and a different approvals process then operates. However, these processes are not specifically undertaken under the *Water Management Act 2000*. If a development is approved under the *Environmental Planning and Assessment Act 1979* (New South Wales), approval under other acts cannot be refused.

5.4.4 Victoria

The Victorian unconventional gas industry is still in its infancy and with moratoriums in place on exploration and hydraulic fracturing. This is unlikely to change in the short term. Despite recommendations by the Victorian Gas Market Taskforce in October 2013 to the remove the moratorium, it will continue until at least May 2015 and there is a parliamentary enquiry pending (Victorian Government, 2013). Potential unconventional gas resources exist. However, the technical and economic feasibility of them is not well understood.

Regulation of unconventional gas in Victoria is split between the *Mineral Resources (Sustainable Development) Act 1990 (MRSD Act)* and the *Petroleum Act 1998*. The *Mineral Resources (Sustainable Development) Act* governs CSG exploration and development activities, while shale and tight gas are governed under the *Petroleum Act*. No specific regulations exist for well integrity or hydraulic fracturing and risks associated with these activities are managed through conditions on a licence. Potential contamination risks are regulated through the *Environment Protection Act 1970*, including the disposal of coproduced water. At this stage, a licence is required to take groundwater under the *Water Act 1989*.

5.4.5 South Australia

Unconventional gas resources in South Australia are governed by the *Petroleum and Geothermal Energy Act 2000*, which cover the licencing requirements, environmental assessment, and approvals for all activities relating to well construction, hydraulic fracturing, and water management. The act does not include an express right to rake water. However, groundwater extraction requires approval under the licence and impacts to groundwater are addressed through a Statement of Environmental Objective and an Environmental Impact Report (Hoare & Finn, 2014). The South Australia State government released the *Roadmap for Unconventional Gas Projects in South Australia* in December 2012. The Roadmap does not regulate any operations, but it does demonstrate the government's commitment to environmental sustainable development of unconventional gas (DMITRE, 2012).

5.4.6 Northern Territory

In the Northern Territory, unconventional gas is regulated through the *Petroleum Act 1984*, *Petroleum Regulations 1994*, and the *Schedule of Onshore Petroleum Exploration and Production Requirements 2012*. All activities undertaken in relation to shale and tight gas exploration and development are governed by this legislation. However, CSG has not been dealt with yet as there are no known CSG resources in the Territory (DME website). The Northern Territory Government released the 'Report of the Inquiry into Hydraulic Fracturing in the Northern Territory' in February 2015. The report contained a range of recommendations, including that there was no justification for a moratorium on hydraulic fracturing and that environmental risks associated with hydraulic fracturing could be managed effectively through the creation of a robust regulatory regime (NTG, 2015).

5.4.7 Western Australia

Western Australia has large shale gas resources in the Perth and the Canning Basins and consequently their legislative focus is on shale and tight gas rather than CSG. Unconventional gas is regulated under the *Petroleum and Geothermal Energy Resources Act 1967* for well construction and gas development activities, while environmental impacts are regulated through the *Environmental Protection Act 1986*. More recently the government released the *Petroleum and Geothermal Energy Resources (Environment) Regulations 2012*, which was developed to address some of the recommendations by the Hunter (2011) review. As part of a regulatory review in 2014 the Department of Mines and Petroleum release the *Draft – Petroleum and Geothermal Energy Resources (Resources Management and Administration) Regulations 2014* for public comment. The Department of Mines and Petroleum have developed these regulations to provide a risk-based management scheme for the exploration for and production of petroleum (including unconventional gas) and geothermal energy resources (Department of Mines and Petroleum website). The Department of Mines and Petroleum also highlight that in the case of operations relating to the exploration or recovery of petroleum they also ensure work is conducted in accordance with good oilfield practice and are compatible with the optimum long-term recovery of petroleum and geothermal energy resources (Department of Minesand Petroleum website). Like many of the other jurisdictions around Australia there is low groundwater use and limited knowledge of aquifers in prospective areas for unconventional hydrocarbon in the Canning Basin in particular. The Northern Perth Basin is better understood than the Canning or Carnarvon Basins, but there are still many issues relating to a lack of understand of aquifers, in particular shallow–deep connectivity and hydrostratigraphy.

5.5 DISCUSSION OF MANAGEMENT APPROACHES

This section discusses the similarities, differences, and challenges in managing hydrogeological impacts associated with unconventional gas development both in Australia and internationally.

5.5.1 Well integrity, hydraulic fracturing and good oil field practice

Risks associated with short term well integrity and hydraulic fracturing are regulated reasonably consistently across jurisdictions and are typically managed to be as low as reasonably possible. It is common for oil and gas industry standards to refer to the concept of 'good oilfield practice', which is a long held industry concept that means 'all those things that are generally accepted as good and safe in carrying out exploration or recovery operations'. This encompasses all activities, including well construction, hydraulic fracturing, and water management where applicable. Currently, there is some flexibility in the design of the regulatory framework to allow for innovation or optimisation (Manifold, 2010). However, this also allows for different interpretation of the regulations and standards, which means the concept of 'good oilfield practice' and the subsequent application and engineering will vary from site to site and between operators (IESC, 2014). The concept of 'good oilfield practice' also appears to be focussed on safety and minimising gas explosions. However, the extent to which 'good oilfield practice' protects the surrounding groundwater resources or environment is not well defined or even well understood.

Well integrity issues are generally manageable with good practices. However well integrity also requires a shared responsibility between government and industry for the protection of surrounding aquifers, which includes rights of existing and future users of the water resources in terms of quantity and quality. The long term impacts of the unconventional gas industry have the potential to be significant and the government, not industry, will be managing these impacts long after the unconventional gas industry has moved on. Some well integrity failure is inevitable only a question of time (Pearce 2005).

5.5.2 Water management

Risks associated with water take for either supply or depressurisation purposes are managed and regulated differently between states. Some unconventional gas reserves are located close to densely populated areas and have the potential to impact water resources that are already in a mature state of use. This has resulted in an acknowledgement that such developments should be referred to the relevant State Water Act as part of the approvals process, even though primary control over development approval may sit with a State's Petroleum and Gas Act.

A key challenge facing policy makers is how best to manage the potential third party impacts particularly from CSG development, both in terms of impacts on individual users, the environment, and the legacy of impact on the total consumptive pool. The option that allows the easiest avenue for development of unconventional gas would involve acceptance of the risk of third party impacts. An alternative is to allow gas development, but require that developers acquire (i.e. purchase) the required water access entitlement via the market or from unassigned water. The risk with this approach is that the volume of water purchased by developers may be large and water purchases of this volume from traditional water users could have long-term impacts on the industries and associated communities that remain long after the gas development has run its course. At the other extreme, the gas development could be stopped in order to protect third party impacts on individual water users and the wider community. There is no easy choice amongst these options and it is not surprising that this poses a significant challenge for policy makers.

There are some broad principles that apply in the context of water resource planning that need to be considered in the context of managing unconventional gas production and its water take. At a high level, the basic principles are that developers require protected rights to take water. This impacts on third parties, including the environment and needs to be managed over the full duration of the potential for impact, which typically exceeds the timeframes over which assessments and projects are carried out. However, the question arises as to how to best achieve these principles. This section discusses potential issues in relation to the basic principles stated above.

Connectivity

The role of hydraulic connectivity in the distribution of hydrogeological impacts is complex. At a simple level, the distribution of impacts is a function of the magnitude of depressurisation necessary to meet production requirements within the gas field (depressurisations or supply) and the level of connection of the target gas formation with the surrounding environment. The level of connectivity is typically more significant with CSG compared to the deeper unconventional gases due to the close proximity of these resources to regional groundwater resources. The magnitude of any depressurisation is a function of a number of aquifer attributes, principally related to hydraulic conductivity, but not exclusively and is also dependent on connectivity. Hence, considering how connected the target gas formation might be to a surrounding aquifer is not a simple task. Also the process of hydraulic fracturing is complex. Fracture propagation can exceed design criteria (due to operational issues or physical heterogeneity) and in some cases may connect previously unconnected aquifers. This could threaten aquifer water quality, not only through the leakage of formation hydrocarbons and fracturing fluids, but through allowing groundwater of an undesirable quality to discharge to adjacent aquifers. This is, however, not likely to be a common situation in an Australian context, unlike areas of eastern United States of America, for example, where these types of issues have been previously documented and at times sensationalised.

Right to take water

Unconventional gas developers need assurance that their development and investment will not be unduly affected by their ability to access water. There are two main approaches as to how this right to take water can be granted. The first, which is supported through the National Water Initiative is to license the right under specific water legislation. As an alternative, the right to take water can be defined under the relevant Petroleum and Gas Act. New South Wales has opted for the former approach to a degree while Queensland has adopted the latter.

The National Water Initiative was developed as a response to the unsustainable use of Australia's water resources. One of the main tenets of the initiative is the introduction of environmentally sustainable extraction levels. This concept was introduced as a means of managing overused and overallocated water systems, but its use also implies that all water systems should be subject to environmentally sustainable extraction levels. The National Water Initiative further advocated that water access entitlements (licensing) be used within a water planning framework as a means of allocating the consumptive use of water within the environmentally sustainable extraction levels. By default, these entitlements should be defined as a perpetual share of water within a defined consumptive pool for a planned area and should be separate to land title.

However, clause 34 of the initiative recognises that there may be special cases associated with mining and petroleum development where the National Water Initiative may not contain the best policies or measures for management. In this context, the National Water Initiative notes that *'specific project proposals will be assessed according to environmental, economic, and social considerations and that factors specific to resource development projects, such as isolation, relatively short project duration, water quality issues, and obligations to remediate and offset impacts, may require specific management arrangements outside the scope of this Agreement.'* NWC (2010) explicitly stated its position that the interception of water by CSG extraction should be licensed to ensure it is integrated into water sharing processes from inception.

There are two major challenges in licensing water take associated with CSG, in particular within the current framework as they represent significant differences to most other licenced water users:

- the volume of water extracted varies considerably overtime, and
- the quality of water extracted is often not suitable for consumptive purposes and therefore may not have been considered part of the total consumptive pool.

A fundamental concept in sustainable water resource management is that of the Total Consumptive Pool. The notion is that water resources are not limitless and within a particular context there are finite volumes of water available for use. The principle of conservation of mass also dictates that once a volume of water is taken from the existing pool, a compensatory impact will occur somewhere else within the pool. The National Water Initiative is very clear that water resource systems should not operate at levels greater than the environmentally sustainable extraction levels.

A water sharing plan (or water management plan), if it is consistent with National Water Initiative, should be based on the concept of deriving the Total Consumptive Pool for a defined area and making entitlement decisions around how best to maximise the benefit of the system. A key technical activity in the planning process is to therefore derive the Total Consumptive Pool volumetrically, which involves an understanding of the boundaries to the water resource subject to the planning activity. While this can be difficult when connectivity across boundaries differs spatially between the target gas formation, aquifers, and surface waters, it is a common hydrogeological issue and is dealt with in most groundwater resource assessments. There may also be significant uncertainty regarding connectivity due to inherent limitations in our ability to precisely determine aquifer connectivity over large potential impacts areas, particularly where conduits (i.e. faults or fractures) are involved.

Essentially, the Total Consumptive Pool assessment is an assessment of the water balance over a long enough period that climate variability has been included. At present, a Total Consumptive Pool analysis is generally restricted to the fresh water resource (hence unlikely to include water held in target gas formations). Within the context of a CSG development, this basic assumption no longer holds. For instance, it is highly likely that the groundwater contained within the target coal measure was either not included in local water resource plans for the area or if included, was not able to be considered, as it was inconceivable that future scenarios would entail pumping large amounts of water from them. Fundamentally, there are now issues as to whether the Total Consumptive Pool that underpins a water resource plan does in fact include the total resource that is currently being accessed by unconventional gas development areas. If the plans do not include them, then simply providing an entitlement under the planned Total Consumptive Pool does not ensure sustainable management. There is a need for detailed assessment of what resources are included within a water resource plan and whether the impacts on the water balance as a result of CSG development have been explicitly modelled prior to providing entitlement for use. To do otherwise is to undermine the rights of existing groundwater users, including the environment.

There are a number of advantages in licensing CSG water take. With a water right comes a series of obligations that apply to the water user, which are usually defined through the Water Act or the relevant water resource plan. Licensing also makes it easier to account for water take within the water resource plan framework, which is particularly relevant where there are resource limits (such as the Sustainable Diversion Limits in the Murray Darling Basin). However, there is currently much debate over whether water held in coal measures is considered part of the Total Consumptive Pool and this will differ between unconventional gas fields, depending on the connectivity with the surrounding aquifers. Licensing can also be a tool in managing third party impacts. The following section discusses this third party impacts management in more detail.

It is worth noting that New South Wales' aquifer interference policy goes a long way to recognising issues around the Total Consumptive Pool definition and management of third party impacts. However, while the Policy requires consideration of these issues and places a large onus on the proponent to provide this information, it does not specify what approach or quality of underlying data is appropriate for such licencing decisions. If policies for this and other jurisdictions were more prescriptive in this regard it would assist.

Managing third party impacts

Unconventional gas fields in Australia are usually located in hydrogeological systems where there are no other water users due to the difficulty in extracting water and the water quality (a notable exception to this is the Walloon Coal Measures in the Surat Basin). However, there is the risk of impacting surrounding water users and environment in neighbouring systems (both surface water and groundwater). The extent to which this occurs will depend on the level of connectivity. Traditionally, third parties have been protected by the derivation of sustainable extraction limits, which protect the ongoing use of the resource. By issuing an entitlement, this conveys certain rights that cannot be reduced to unacceptable levels. An ill-defined part of this process is a twofold question: what rights are conveyed to a water user by the declaration of a sustainable extraction limit within a water resource plan and what level of impact is considered unacceptable within a groundwater context (and how is it quantified)?

Consideration of third party impacts is discussed further in the following sections within the context of the level of connectivity. Where groundwater extraction or gas development is occurring in an aquifer that is very highly or highly connected to a neighbouring system, the water take will have a direct and immediate impact on water users in that neighbouring system. In this instance, a gas development is not accessing 'new' water, but rather taking water from a neighbouring resource. The impact may be seen in the Total Consumptive Pool or as drawdown. This is recognised in the Queensland planning system with an acceptable level of drawdown documented. However, a core purpose of water resource planning is to establish sharing rules for access to the resource and to ensure protection of existing licence holders, including the environment.

If water extraction as a result of unconventional gas development is licensed, the question then arises as to how best to secure access to this water. If there is unassigned water available in the resource, this water could be made available for gas development. In this instance, a policy decision is required around how best to allocate this water. In some regions, the water is allocated on a 'first in' basis. An alternative, which may be preferable given the size of the resource, would be to conduct an auction for the unassigned water.

In systems where there is no unassigned water (i.e. the system is fully allocated), it is likely that a water resource plan would require unconventional gas operators to purchase water entitlements from existing water users as would be the case for other new developments. The price of the water would be established by the water market. An alternative to permanent purchase of water rights would be to create a leased water product. The water market would again establish a price for this product. In either case, there is an advantage to these approaches in requiring proponents to make purchases on the market ensures that existing users are only affected should they choose to sell their water. The price at which they choose to sell is also their choice. A challenge will be the size of the gas operation relative to other industries and the likelihood that the CSG industry will be able to pay higher prices than other water users. While this would in fact move water to the higher value use if water moves entirely (or predominantly) towards the onshore gas industry there may be flow-on effects within the broader community as other water supported industries become limited. Flow-on effects for dependent industries may also manifest and these secondary effects are harder to manage within the market. Once the gas operation is complete, it may be hard to rebuild these (and dependent) industries. Within this broader context it may also be appropriate to quarantine part of the water resource for specific use (either consumptive or non-consumptive) on top of environmental water requirements.

If CSG water extraction is not licensed, third party impacts are instead managed through 'make good' arrangements. The difficulty with this approach is that individual land owners do not have the option to enter into the agreement. The impact on their right to water is already given and the negotiation is only over price. The established price in this instance is effectively a 'mock' market fee, rather than using the water market to reflect the true value of the water right.

Understanding the impact of unconventional gas water use on other water users becomes complex when the impacts are experienced over the short to medium term. In this instance, while there will clearly be an additional third party impact from the unconventional gas water use, understanding the likely volume and timing of this impact is challenging. The impact on the target gas formation will be immediate. However, the impact on a neighbouring or overlying aquifer system may be incremental over long periods. Such impacts would need to be considered in the water accounting system established, particularly if there is an extraction limit in place in the neighbouring or overlying aquifer system.

Systems with low and very low connectivity

In systems that have low impact and long delays to impacts it is easier to understand how unconventional gas water use could be managed through a separate process to the traditional licensing and water resource planning. However, it is worth considering that while there may be a significant time lag for impacts (greater than 50 years) the impact may still be significant in terms of the overall water balance or contamination. It is difficult to see how this would be considered within the water resource planning framework, as the time horizon is substantially longer than the usual planning horizon. These legacy effects, though, will need to be managed and most likely by government once the CSG operation has been completed and future water users will be impacted through no fault of their own. In terms of environmental impacts, there is similar potential for legacy issues in some cases.

5.6 CONCLUSIONS

There are a large number of regulations that control the unconventional gas industry, both in Australia and worldwide. While there are different hydrogeological risks associated with different types of unconventional gas, there are also many common elements. There are currently three states producing unconventional gas, shale gas in Cooper Basin in South Australia, and coal seam gas in Queensland and New South Wales, with a number of other states in exploration and project scoping phases. Each jurisdiction manages the hydrogeological risks differently. There are many similarities in the way risks associated with well integrity and hydraulic fracturing are regulated and managed. However, there are significant differences in the way water access rights and third party impacts are managed and regulated. There are compelling reasons why water take in association with gas development needs to be incorporated into the water resource planning framework, so there is clear and accountable water use. This is the most transparent way to ensure that unconventional gas development can occur whilst still protecting the rights of other water users and the environment.

However, in many cases, water take does not fit neatly into the existing licensing framework, which needs to rapidly adapt to this critical issue. Further work is required to understand if it is achievable to make water take in association with unconventional gas development subject to an entitlement licence within a water resource plan, which are typically limited to a short (20 year) planning period. Many impacts relating to unconventional hydrocarbon resources may not become apparent for longer time frames. Well integrity failure legacy issues due to aggressive aquifer water and seismic activity need to be better incorporated in all jurisdictions.

The time frames for managing impacts from CSG water take will vary between fields and is a complex function of many variables. The only certainty is that the time frame will be such, that it will be outside the time frame of the gas development. It will therefore be managed by government as a legacy issue long after the gas industry has moved on. It is therefore imperative that there is a shared responsibility between government and industry for the protection of the rights of existing and future users of the water resources (including the environment) in terms of quantity and quality. A licensing framework provides a clear and transparent means for accounting for water use. However, exactly how CSG water take is incorporated in the water allocation framework is likely to require a flexible approach that allows for different scenarios, depending on the level of connectivity between target gas formations and surrounding water resources. There are still significant challenges for policy makers with no clear cut readymade solutions. It is hoped that by highlighting the issues in a review such as this, policy makers can be better informed and subsequently make the best possible policies, minimising impacts, while maximising commercial opportunities for all interest groups.

REFERENCES

API (2009) *Hydraulic Fracturing Operations – Well Construction and Integrity Guidelines*, American Petroleum Institute: 1st ed., Washington, [Online] Available from: http://www.api.org/~/media/Files/ Policy/Exploration/API_HF1.ashx [May 2015]

Birgit, V., Aadnoy, B.S. (2008) *Well-Integrity Issues Offshore Norway*. SPE/IADC 112535. Conference presentation, SPE/IADC Drilling Conference 2008. 4–6 March, 2008. Orlando, USA.

Bourgoyne, Jr. A.T., Scott, S.L., Manowski, W. (1999) *A Review of Sustained Casing Pressure Occurring on the OCS*. US. Department of Interior, Washington, D.C. [Online] Available from: http://www.wellintegrity.net/Documents/MMS%20-%20Review%20of%20SCP%20-%202000. pdf [May 2015]

Cook, P., Beck, V., Brereton, D., Clark, R., Fisher, B., Kentish, S., Toomey, J., Williams, J. (2013) *Engineering energy: unconventional gas production*. Australian Council of Learned Academies.

Corneliussen, K., Sørli, F., Brandanger, H., Tenold, E., Menezes, C., Grimbert, B. (2007) *Well Integrity Management System (WIMS) – A Systematic Way of Describing the Actual and Historic Integrity Status of Operational Wells*. SPE Annual Technical Conference and Exhibition. November 2007. Anaheim, USA.

CSIRO (2012) *Fact sheet – Australia's shale gas resources*. [Online] Available from: http://www.csiro. au/Outcomes/Energy/Energy-from-oil-and- gas/UnconventionalGas/Learn-more/Australias-shale-gas-resources.aspx [May 2015]

Davies, R.J., Almond, S., Ward, R.S., Jackson, R.B., Adams, C., Worrall, F., Herringshaw, L.G., Gluyas, J.G., Whitehead, M.A. (2014) *Oil and gas wells and their integrity: Implications for shale and unconventional resource exploitation*. Marine and Petroleum Geology 2014, pp. 1–16 [Online] Available from: http://dx.doi.org/10.1016/j.marpetgeo.2014.03.001 [May 2015]

Department of Primary Industries (2012) *New South Wales Aquifer Interference Policy*. Office of Water. [Online] Available from: http://www.water.NewSouthWales.gov.au/__data/ assets/pdf_file/0004/549175/NewSouthWales_aquifer_interference_policy.pdf [May 2015]

DME (n.d.) *Unconventional Oil and Gas*, Department of Mines and Energy [Online] Available from: http://www.nt.gov.au/d/Minerals_Energy/index.cfm?header=Unconventional %20Oil%20and%20Gas [May 2015]

DMITRE (2012) *Roadmap for Unconventional Gas Projects in South Australia*. Department for Manufacturing, Innovation, Trade, Resources and Energy – Energy Resources Division. [Online] Available from: http://www.petroleum.dmitre.sa.gov.au/__data/assets/pdf_file/ 0008/179621/Roadmap_Unconventional _Gas_Projects_SA_12-12-12_web.pdf [May 2015]

DMP (2013) Petroluem Fact Sheet. Western Australian Government Department of Mines and Petroleum. [Online] Available from: http://www.dmp.wa.gov.au/documents/ 132499_Resources_Type_Fact_Sheet.pdf [May 2015]

DNRM (2013) *Code of Practice for constructing and abandoning coal seam gas wells and associated bores in Queensland*. Queensland Government, Department of Natural Resources and Mines, Edition 2, September 2013. [Online] Available from: https://www.dnrm. qld.gov.au/__data/assets/pdf_file/0011/119666/code-of-practice-csg-wells-and-bores.pdf [May 2015]

DoE (2015) *Great Artesian Basin Sustainability Initiative.* Commonwealth Department of the Environment. Available from: http://www.environment.gov.au/water/environment/great-artesian-basin-sustainability-initiative [May 2015]

EDO (2011) *Mining Law in New South Wales Discussion Paper.* Environmental Defender's Office NEW SOUTH WALES June 2011. [Online] Available from: http://www.edoNew SouthWales.org.au/mining_law_in_new_south_wales_discussion_paper_june_2011 [May 2015]

Frogtech (2013) *Geological risks of shale gas in Australia.* January 2013. Report for the Australian Council of Learned Academies, Securing Australia's Future: Project Six Engineering Energy: Unconventional Gas Production. [Online] Available from: http://www.acola.org.au/PDF/SAF06FINAL/Frogtech_Shale_Gas_ Geology_and_Risks%20Jan2013.pdf [May 2015]

Gas Today (2007) *Tapping into Australia's tight gas.* Gas Today, November 2007. [Online] Available from: http://gastoday.com.au/news/tapping_in_to_australias_tight_gas/4531 [May 2015]

Geoscience Australia (2010) *Australia's Identifiable Mineral Resources* (AIMR). [Online] Available from: http://www.ga.gov.au/scientific-topics/minerals/mineral-resources/aimr [May 2015]

Geoscience Australia and BREE (2014) *Australian Energy Resource Assessment* 2nd Ed. Geoscience Australia, Canberra. [Online] Available from: http://www.ga.gov.au/corporate_data/79675/79675_AERA.pdf

Geoscience Australia (2015) *Coal seam gas.* [Online] Available from: http://www.ga.gov.au/scientific-topics/energy/resources/petroleum-resources/coal-seam-gas [May 2015]

GISERA (2011) *Frequently asked questions on coal seam gas extraction and fraccing.* Gas Industry Social and Environmental Research Alliance. [Online] Available from: http://www.gisera.org.au/publications/faq/faq-csg-extraction-fraccing.pdf [May 2015]

Hoare, R., Finn, W. (2014) *Regulation of Hydraulic Fracturing in Australia.* Australia Water Association, OzWater Conferenced 2014. [Online] Available from: http://www.awa.asn.au/htmlemails/Ozwater14/pdf/ 40740000Final00328.pdf [May 2015]

Hunter, T. (2011) *Shale, coal seam and tight gas activities in Western Australia – Analysis of the capacity of the Petroleum and Geothermal Energy Act 1967 (WA) to regulate onshore gas activities in Western Australia.* Law papers (2011): 1–28. [Online] Available from: http://works.bepress.com/tina_hunter/45 [May 2015]

IESC (2014) *Bore Integrity – Background Review.* Published by the Independent Expert Scientific Committee on Coal Seam Gas and Large Coal Mining Development. Department of the Environment. June 2014. [Online] Available from: http://www.environment.gov.au/system/files/resources/00f77463-2481-4fe8-934b-9a496dbf3a06/files/background-review-bore-integrity.pdf [May 2015]

King, G. (2012) *Hydraulic Fracturing 101: What Every Representative, Environmentalist, Regulator, Reporter, Investor, University Researcher, Neighbour and Engineer Should Know About Estimating Frac Risk and Improving Frac Performance in Unconventional Gas and Oil Wells.* Society of Petroleum Engineers. [Online] Available from: https://www.onepetro.org/conference-paper/SPE-152596-MS [May 2015]

Manifold, C. (2010) *Why is Well Integrity Good Business Practice?* Petroleum Exploration Society of Australia (PESA) News June/July 2010.

NGT (2015) *Report of the Independent Inquiry into Hydraulic Fracturing in the Northern Territory.* Northern Territory Government. February 2015. [Online] Available from: http://www.hydraulicfracturinginquiry. nt.gov.au/index.html [May 2015]

New South Wales Trade and Investment (2012a) *Code of Practice for Coal Seam Gas Fracture Stimulation Activities.* New South Wales Department of Trade and Investment, Regional Infrastructure and Services, Resources and Energy. September

2012. [Online] Available from: https://www.NewSouthWales.gov.au/sites/default/files/csg-fracturestimulation_sd_v01.pdf [May 2015]

New South Wales Trade and Investment (2012b) *Code of Practice for Coal Seam Gas Well Integrity*. New South Wales Department of Trade and Investment, Regional Infrastructure and Services, Resources and Energy. September 2012. [Online] Available from: http://www.resourcesandenergy.NewSouthWales.gov.au/__data/assets/pdf_file/0006/516174/Code-of-Practice-for-Coal-Seam-Gas-Well-Integrity.PDF [May 2015]

NWC (2010) Position Statement: *The coal seam gas and water challenge*. National Water Commission. [Online] Available from: http://nwc.gov.au/__data/assets/pdf_file/0003/9723/Coal_Seam_Gas.pdf [May 2015]

Nygaard, R. (2010) *Well Design and Well Integrity, Wabamum, Areas CO₂ Sequestration Project (WAOGIASP)*. Energy and Environmental Systems Group. Institute for Sustainable Energy, Environment and Economy (ISEEE). [Online] Available from: https://www.ucalgary.ca/wasp/Well%20Integrity%20Analysis.pdf [May 2015]

QWC (2012) *Underground Water Impact Report – Surat Cumulative Management Area*. Queensland Water Commission. [Online] Available from: https://www.dnrm.qld.gov.au/__data/assets/pdf_file/0016/31327/ underground-water-impact-report.pdf [May 2015]

Royal Society and the Royal Academy of Engineering (2012) *Shale gas extraction in the UK: a review of hydraulic fracturing*. [Online] Available from: http://www.raeng.org.uk/publications/reports/shale-gas-extraction-in-the-uk [May 2015]

SCER (2013) National Harmonised Regulatory Framework for Natural Gas from Coal Seams. Standing Council on Energy and Resources. [Online] Available from: http://scer.govspace.gov.au/files/2013/06/National-Harmonised-Regulatory-Framework-for-Natural-Gas-from-Coal-Seams.pdf [May 2015]

SKM (2012) A Leading Practice Framework for Coal Seam Gas Development in Australia. Advisory report for the Department of Resources, Energy and Tourism prepared by Sinclair Knight Merz. Unpublished report.

SKM (2013) *Unconventional gas in Australia – Infrastructure needs*. Report for the Australian Council of Learned Academies, Securing Australia's Future: Project Six Engineering Energy: Unconventional Gas Production, Dr Richard Lewis, SKM. Unpublished report.

US EPA (2012) *Plan to study the potential impacts of hydraulic fracturing on drinking water resources: status update*. United States Environmental Protection Agency. Report number EPA 601/R-12/011. [Online] Available from: www.epa.gov/hfstudy [May 2015]

Victorian Government (2013) *Gas Market Taskforce – Final report and recommendations*. [Online] Available from: http://www.energyandresources.vic.gov.au/about-us/publications/Gas-Market-Taskforce-report [May 2015]

Watson, T., Bachu S. (2009) *Evaluation of the Potential for Gas and CO₂ Leakage Along Wellbores*. Society of Petroleum Engineers. [Online] Available from: https://www.onepetro.org/journal-paper/SPE-106817-PA [May 2015]

Yeo, B. (2012) *Wraps: Oil and Gas. Australian shale and tight gas hits milestone as Norwest Energy, Exoma Energy increase activity*. Proactive Investors. [Online] Available from: http://www.proactiveinvestors.com.au/companies/news/32494/australian-shale-and-tight-gas-hits-milestone-as-norwest-energy-exoma-energy-increase-activity-32494.html [May 2015]

Chapter 6

Legacy pesticide contamination in Aarhus – groundwater protection and management

N.C. Pedersen[1], E. Stubsgaard[1], L. Thorling[2], R. Thomsen[2],
V. Søndergaard[2] & B. Vægter[3]

[1]Department of Agriculture and Water Management, Municipality of Aarhus, Denmark
[2]Department of Groundwater and Quaternary Geology Mapping, Geological Survey of Denmark and Greenland, Aarhus, Denmark
[3]Aarhus Water Ltd., Aarhus, Denmark

ABSTRACT

In Denmark, a range of regulatory instruments are available to counter pesticide pollution. The statutory provisions are primarily based on voluntary measures, which may be supplemented by injunctions if this is considered necessary in specific cases. Targeted regulatory intervention such as injunctions to achieve pesticide-free production is required due to the discovery of extensive pesticide contamination of the groundwater throughout the country. Analysis of some 20 years of monitoring data has shown pesticides in about every third well, with the drinking water threshold being exceeded in about one in every six wells.

Detailed mapping of the hydrogeological/geochemical status of aquifers was undertaken in Aarhus, facilitating identification of the areas where the risk of pesticide contamination is greatest, i.e. the areas where supplementary regulatory efforts are warranted.

Since 1999, information campaigns have been implemented in these areas and farmers have been offered compensation for pesticide-free production. This voluntary scheme has only had limited effect and since 2013 it has been supplemented by a possibility of imposing pesticide-free production in groundwater protection zones (vulnerable areas). The costs are primarily compensations given to the farmers who convert to pesticide-free production and are funded collaboratively by the water service providers. Costs for voluntary measures and injunctions will be equivalent to EUR0.07 per m^3 abstracted water over the next 20 years. Additionally, all publicly owned areas are kept pesticide-free and the authorities have initiated measures targeting historical point sources.

6.1 INTRODUCTION AND AIM

With a total of 300 000 inhabitants the Municipality of Aarhus is the second largest in Denmark. As in the remaining parts of Denmark, groundwater is the only drinking water source and the production of drinking water is based exclusively on groundwater treated through aeration and filtering. Groundwater abstraction in the area has been intensive over the past 50 years with large capacity abstraction wells distributed all over the Municipality. Since the 1980s, Aarhus Water and the County of Aarhus have collaborated on modelling and mapping efforts to provide data to help them manage groundwater abstraction in Aarhus Municipality. The groundwater resources in Aarhus Municipality are described as critical, keeping in mind current water consumption and available sources. A high level of groundwater protection is therefore needed to ensure that future needs for drinking water are covered (Municipality of Aarhus, 2010).

The majority of Aarhus Municipality is classified as high priority for drinking water protection. The mean groundwater recharge flux in the deep aquifers used for drinking water is 50 mm annually. Hence, the total annual recharge for the aquifer is 25 Mm^3. A considerable number of water abstraction wells, primarily located in rural areas, supply the municipality with some 20 Mm^3 of drinking water annually. This represents about 80% of total recharge (Municipality of Aarhus, 2010).

The aim of this chapter is to document the challenges that scientists and regulators have had to overcome to address these issues.

6.2   DANISH LEGISLATION

Danish environmental policy is based on the principle of prevention and on implementing counter measures at source. This means that the Danish groundwater resource shall be safeguarded against pollution and that the preventive efforts made to avoid groundwater pollution shall be given a higher priority than the subsequent treatment of polluted groundwater (Miljøministeriet, 2010).

In Denmark, water supply is entirely based on groundwater abstraction. Management intervention was obviously required in the 1990s, when pesticide monitoring of groundwater was introduced in Denmark. It quickly became clear that the groundwater was contaminated at many locations. In 1994, the Danish Government introduced a 10-point plan for future protection of groundwater. The 10-point plan was implemented in a new Groundwater Protection Act adopted by the Danish Parliament in 1998. In pursuance of this act 'high priority water abstraction areas' were designated in all regions of Denmark. In its current form the aim was to designate large cohesive areas, so that significant portions of the regional demand for drinking water could be covered and entire watersheds included. These areas cover approximately 35% of Denmark (GEUS, 2015a).

6.2.1   Groundwater protection plans

The Environmental Protection Act (Miljøministeriet, 2010) and the Water Supply Act (Miljøministeriet, 2013) are the main pillars of the legislative framework on groundwater protection in Denmark. The focus of the Environmental Protection Act is the prevention of contamination and specific requirements for protection of the groundwater may be introduced as required in pursuance of this act. The main objectives of the Water Supply Act are to ensure the planned use of groundwater resources and that the nation's water supply is adequate with respect to both quality and quantity.

In accordance with the Water Supply Act, the Aarhus municipality must adopt Groundwater Protection Plans for all high priority water areas. These plans identify potential groundwater contamination sources and describe the measures needed to safeguard the groundwater from contamination. The measures may be either specific actions, targeting known pollution sources in the area or guidelines specifying how the municipality will process applications for activities that may pollute the ground-water. The protection plan is evaluated through a public planning process, which ensures a high degree of transparency and public participation. Moreover, the protection plan shall include an implementation timetable and shall identify those responsible for implementing the plan.

Through the 1998 Act, the municipalities were given the authority to implement the necessary restrictions. In groundwater protection zones it may be necessary to stop or limit the use of pesticides and a wide range of supplementary instruments are available for this purpose, e.g. subsidies for afforestation, cultivation agreements with farmers, introduction of organic farming, or state acquisition of the areas with a view to afforestation. Some other pertinent points are:

- the water service providers hold a central role in the efforts made to reach a voluntary agreement with each farmer concerning the necessary measures,
- supplementary to the voluntary measures, the 1998 Act introduced the opportunity to impose mandatory restrictions on the use of for example pesticides,
- any loss suffered by the landowner due to any voluntary or mandatory restrictions shall be compensated in full by the water service providers, and
- the act only authorizes restrictions on commercial use. Consequently, private persons' use of pesticides cannot be limited or banned.

6.2.2 Wellfield protection zones

In pursuance of the Environmental Protection Act, well field protection zones have been created. They serve to protect wells against pesticide contamination among others (see Section 6.3). A physical protection area with a 10 m radius around individual supply wells is stipulated in the Environmental Protection Act. Within this area, only supply-related activities are permitted and the use of pesticides is banned. Furthermore, in a 25 m zone surrounding the wells land shall not be farmed, nor may pesticides be used.

In 2007, the Danish Ministry of the Environment introduced another zone. The size of the zone depends on the size of the abstraction and the geological conditions, and in Aarhus Municipality the zone normally has a 25–200 m radius.

Within this zone the use of pesticides can be banned if there is a potential risk of contaminating the well. A distinctive characteristic of this zone is that *any* use of pesticides can be banned, both for professional and private purposes, e.g. private gardening.

6.3 NATIONAL DANISH MAPPING, PROTECTION AND MONITORING OF THE GROUNDWATER

National planning and implementation of groundwater protection in Denmark is based on detailed knowledge about the groundwater resources and any potential sources of groundwater pollution. The continuous monitoring of the groundwater at the supply/treatment facility, in well fields, and at other select locations has provided considerable understanding of groundwater resources. There have been considerable amounts of groundwater investigation in Denmark.

The results of detailed hydrogeological mapping throughout the 1980s and 1990s in Aarhus formed the basis for the Danish Parliament's decision to launch an ambitious plan for national hydrogeological mapping. Since 1998, the aquifers and the capacity of the overlying layers for natural protection of the groundwater against various forms of pollution have been mapped in detail in accordance with the Parliament's decision. This project (Thomsen, R. & Sondergaard, V. (Tech. Ed.) & Klee, P. (Ed.) (2013)) has been a three-step process consisting of:

- Spatially dense hydrogeological mapping based on existing data and supplemented with new geophysical surveys, survey wells, water sampling, hydrological modelling, etc. aimed at facilitating the establishment of site-specific protection zones. The protection zones are to be established on the basis of model calculations of groundwater flow and calculations of the degradation of the contamination from point sources and diffuse sources, taking into account knowledge of the local geochemical conditions.
- Mapping and assessment of all past, present, and possible future sources of contamination—both point and diffuse sources.
- Preparation and evaluation of a groundwater protection plan (see Section 6.2.1 Groundwater protection plans), stipulating politically determined regulations for future land use within the site-specific groundwater protection zones.

These mapping efforts will be concluded by the end of 2015 and maps will then cover all the high priority water abstraction areas, i.e. about 35% of the country. The total costs of the mapping and planning of measures is EUR350 000 000. The groundwater mapping project has captured the distribution of aquifers and their connection with shallow soils, producing groundwater vulnerability maps. These maps also display the site-specific groundwater protection zones (Thomsen *et al.*, 2013).

6.3.1   Knowledge-based administration and knowledge sharing

In Denmark, municipal administration of the groundwater resource is based on the ideal of knowledge-based administration and knowledge sharing. All groundwater data are available in public databases and may be used freely by any citizen and by the authorities (GEUS, 2015b). This applies to all drilling/lithology data, water analyses from the government groundwater monitoring initiative, the auto-monitoring scheme at the supply/treatment facility, data from geophysical studies, and well data from private and public water abstraction wells. Reporting of well information (water levels and quality) is mandatory for all abstraction wells for irrigation. A national groundwater monitoring programme (GEUS, 2015c) was initiated in 1988. Since then the groundwater has been monitored at about 1000 monitoring points representing Denmark's hydrogeological variability. In addition to common ions, such as nitrate, chloride, and sulphate, a wide range of trace elements such as nickel and arsenic are also measured. It also includes a dynamic pesticides analysis programme that is continuously aligned with current knowledge on the mobility and degradation of pesticides and a range of anthropogenic substances including chlorinated solvents, phthalates and BTEX.

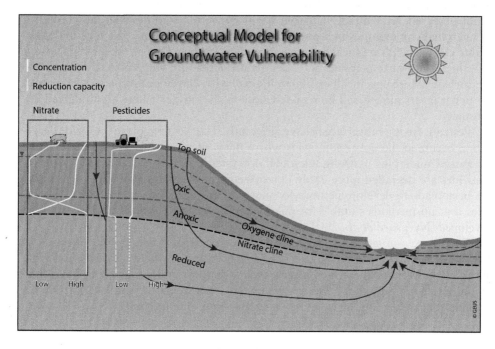

Figure 6.1 Conceptual model for the vulnerability of the groundwater to nitrate and pesticides. The white and yellow lines in the boxes illustrate the development with depth in concentration and reduction capacity for nitrate and pesticides.

Groundwater monitoring provides a basis for measuring the benefit of national environmental/groundwater initiatives and has considerably increased our general knowledge about the condition of groundwater. Knowledge gained from the groundwater monitoring initiative informs hydrogeological conceptual models, in particular those relating to groundwater vulnerability in Denmark (Hansen *et al.*, 2011).

6.3.2 National groundwater vulnerability

In order to achieve site-specific protection of groundwater against pollution, a defendable conceptual model, incorporating a site-specific understanding of groundwater vulnerability is essential. The need for conceptual models has even been described in the European Union Groundwater Directive (EU, 2006). Nitrate and pesticides are the two primary sources of pollution in the open land where the majority of the abstraction catchment areas for the water abstraction wells are located.

Figure 6.1 presents a conceptual model for leaching of nitrate and pesticides. Both types of pollution may be degraded through processes that occur naturally in the environment, but the solute transport and degradation dynamics of the two types of pollution are fundamentally different. Therefore, the vulnerability of the groundwater with respect to each type of pollution also varies.

Nitrate (from fertilisers, animal waste, septic systems, etc.) can leach through the soil and persist in the groundwater if oxic conditions are present in the aquifers.

Nitrate disappears at a certain depth, the so-called 'nitrate cline', below which nitrate is not observed, as available nitrogen but is converted to ammonia and consumed by microbial communities in a process known as denitrification. The location of the nitrate cline depends on local hydrogeological conditions. One of the most important tasks in groundwater mapping is to identify areas that are vulnerable to nitrate from the surface. More specifically, if laterally extensive clayey sediments are present in the upper layers nitrate will be retained close to the surface where it will degrade/be consumed.

Pesticides in the subsurface behave differently. The most important pesticide degradation occurs in the upper soil layers where the content of readily degraded organic substances underpins a strong microbial ecosystem. The overwhelming majority of pesticides are degraded more easily in oxidised conditions. In Denmark, an attempt has been made to develop methods for mapping differences in vulnerability to pesticides, but no methods suitable for public administration purposes have so far been developed (Nyegaard et al., 2004).

Therefore, estimation of vulnerability to pesticides is based on identification of areas where groundwater recharge is particularly large and where the risk of pollution of the groundwater is therefore higher than at other locations. Areas with substantial groundwater infiltration are also frequently the areas where the soil layers above the aquifers are sandier and, hence more vulnerable to nitrate leaching.

6.4   MAPPING AND PROTECTION OF THE GROUNDWATER IN AARHUS MUNICIPALITY

The population of Aarhus increased considerably throughout the 1960s and increased abstraction of groundwater caused a considerable drop in the groundwater level with increasing sulphate concentrations at several well fields. Towards the end of the 1980s, the former Aarhus County and Aarhus Water (local water supply provider) agreed to collaborate in preparing a numerical groundwater model to be able to determine more precisely which level of abstraction is sustainable at all well fields within the municipality.

The results of the model dictated that the regional limits for groundwater abstraction had to be adjusted from 27 Mm^3/year down to 22 Mm^3/year, which occurred in 1993. Abstraction at the Beder well field (one of the major well fields for public water supply) was reduced from 5.5 Mm^3/year to 3.5 Mm^3/year.

This measure was a clear benefit to the groundwater resource with the previously declining groundwater levels, recovering 8 m over the next 15 years (Figure 6.2). Concurrently, with the reduction of the abstraction water quality improved with stabilising sulphate content (possibly even beginning to decline) as groundwater levels recovered.

The hydrogeological modelling process also showed that current knowledge about the spatial distribution of the recharge areas for the aquifers was inadequate for the precise calculation of a sustainable abstraction. Aarhus County and Aarhus Supply Services decided to collaborate on a local-scale detailed mapping of the recharge windows for aquifers and their vulnerability to contamination. Interpretation of model results also enabled the authorities to conclude that the total potential for abstraction in the municipality was nearly exploited.

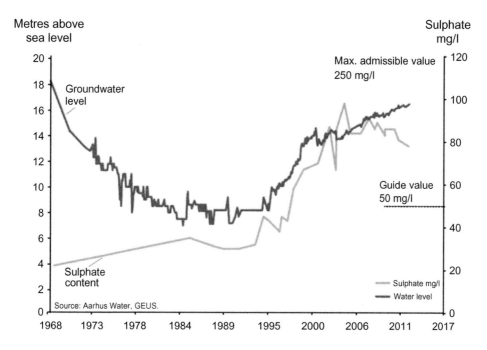

Figure 6.2 Decline and rise of groundwater level in response to groundwater abstraction in Beder and the resulting rise in sulphate concentration in the groundwater.

In the early 1990s, Aarhus University developed geophysical methods that proved to be particularly well suited for these mapping activities. Thomsen (2004; 2007) described the application and results of these geophysical techniques in and around Aarhus in the 1994–1997 period. They significantly improved the traditional hydrogeological mapping based exclusively on borehole information. The improvements were achieved by using borehole information to better calibrate geophysical measurements and to provide a well-documented, detailed, and useful basis for area-specific identification of groundwater protection zones.

In 1998, mapping of the extent of the aquifers was concluded and the first ever detailed vulnerability map covering the municipality was published. The vulnerability map was based on the total thickness of clay layers above the aquifers as its thickness was proven to be of considerable importance in determining recharge and the protection of groundwater quality across Denmark.

Groundwater protection zones are now being established in the high priority water abstraction area on the basis of spatially dense geophysical mapping. The location of the protection zones is used in physical planning in Aarhus and new urban developments around Aarhus city will usually not be permitted in areas where natural protection of the groundwater is poor. Location of other potentially polluting activities is either banned or restricted in the groundwater protection zones.

Mapping of the high priority areas will be completed in Aarhus by the end of 2015. The area of Aarhus Municipality is 468 km² and approximately 310 km² (66%) of it is

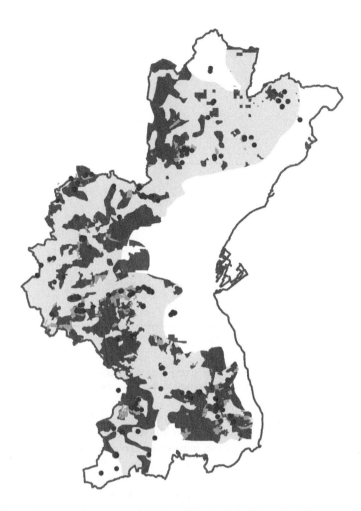

Figure 6.3 Groundwater protection zones (vulnerable areas) (red areas) within the highpriority area
for drinking water protection (light blue areas). Groundwater protection zones protected
in 2015 through voluntary agreements on pesticide-free production are presented in light
green. The water service providers drinking water wells are shown as blue dots.

designated as a high priority water abstraction area. The current mapping has revealed
that 121 km² are vulnerable (groundwater protection zones), which constitutes 25%
of the municipality (Figure 6.3).

6.5   AARHUS CONTAMINATION MITIGATION MEASURES

Since 1999 Aarhus Water, which provides approx. 85% of the water consumed in
Aarhus has focused on minimising the risk of pesticide contamination by implementing
information campaigns and by entering into agreements on pesticide-free production.

Until 2013, the agreements were voluntary for farmers, but in 2013, Aarhus Municipality decided to introduce injunctions against farmers who would not voluntarily comply. Twenty-four of the 25 municipal water service companies joined forces to secure funding for this measure. In relevant urban areas, pesticide contamination prevention efforts are still focused on information campaigns.

6.5.1 Voluntary agreements process

In an effort to minimise pesticide pollution of the groundwater, Aarhus Water has run groundwater protection campaigns in the 1999–2013 period. A key element of the campaigns consists of offering farmers voluntary agreements to undertake pesticide-free production within 300 m zones surrounding abstraction wells and in those parts of the groundwater catchment areas that have been identified as groundwater protection zones, as described previously. Farmers were offered either perpetual agreements or agreements with a 5, 10, or 20 year validity period, which can subsequently be mutually extended.

All landowners with more than 3 ha of land in the campaign areas were contacted by letter and phone and offered a meeting at their own premises. At the meeting the landowners were informed about how groundwater recharge occurred and about the consequences that field spraying can have for groundwater quality.

Production agreements were based on a collaboratively developed replacement model negotiated between the national water works associations and the farmers' organisations (The National Agreement, DANVA (2000)).

The agreements compensate for the production loss suffered by producers. In addition to provisions, stipulating how calculations of compensation shall be made when each of the various restrictions were introduced, The National Agreement also included a draft standard agreement, a description of the legal basis and the proposed restrictions, and finally a background report prepared by the farmers' research institution. The agreements options include pesticide-free production, reduced use of nitrogen, no pesticides plus reduced use of nitrogen, and finally permanent grassing.

6.5.2 Information campaigns

During the information campaigns it became clear that many conventional farmers were sceptical about entering into agreements on pesticide-free production. Therefore, to improve the uptake of voluntary agreements, Aarhus Water offered farmers an advisory service on pesticide-free farming. Farmers were offered advice on organic farming and alternative conventional production where robust crops and mechanical weed control is used. Furthermore, interested farmers were put in touch with organic farmers in the area with an interest in leasing the land.

The demand for energy crops has opened up new alternative crops for conventional farmers because these are often perennial crops and can be produced without use of pesticides and do not require any special organic farming expertise. In addition private afforestation may be an alternative, typically in more limited areas, as landowners may combine an agreement on pesticide-free production with national financial support for afforestation.

The farming information campaigns implemented in the 1999–2013 period were a success in as much as they contributed to disseminate essential information to farmers on the importance of focusing on minimising the risk of pesticide pollution, but the success rate of the voluntary agreements was relatively modest. From 1999 to 2013, a total of 1000 ha of farming land in 300 m zones surrounding wells and in the groundwater protection zones were protected through voluntary agreements on pesticide-free farming. After 13 years of efforts only 1/6 of the target area had been protected.

6.5.3   Injunctions to achieve pesticide-free production

In 2013, due to the lack of uptake of voluntary agreements, Aarhus Municipality started preparing a new generation of groundwater protection plans (Aarhus Kommune, 2013). According to these plans the municipality may impose restrictions concerning pesticide-free production on landowners in groundwater protection zones and near wells amongst other areas. In its plans the municipality invites the water works to continue their information campaigns as previously and encourages water works to continue offering voluntary agreements with landowners. The water works have been granted a 2–5 year period during which they shall attempt to establish voluntary agreements. Subsequently, the municipality will impose restrictions on landowners who have not entered into voluntary agreements in the form of injunctions to initiate pesticide-free production.

2014 saw the first farming campaign based on the new generation of groundwater protection plans (Aarhus Municipality, 2013). A total of 105 landowners who had areas within groundwater protection zones or which were located within the near-well protection areas were contacted. As in previous campaigns, landowners were offered a meeting where they were informed that they had the opportunity to enter into a voluntary agreement with the water works in Aarhus.

By the beginning of 2015, a total of one third of the landowners had agreed to enter into a voluntary agreement. Mainly smaller recreational farmsteads accepted the agreements. In all, half of the farmers have declined a voluntary agreement. This includes the majority of larger farms who are now awaiting an injunction from the municipality. Injunctions may be imposed 2 years after the plan was adopted, i.e. as from April 2015.

6.5.4   Distribution of land

Distribution of land can be an effective tool in campaign areas where scattered parcels of land need to be protected through production (farming) agreements. In a land distribution process the farmers are given the opportunity to swap, buy, or sell areas. Often only parts of a farmer's areas will have been classified as vulnerable (groundwater protection zones). In the eyes of many farmers production conditions are inopportune when parts of the fields need to be farmed without pesticides, whereas other areas owned by the farmer may still be farmed conventionally, using pesticides. In Denmark, land distribution processes are implemented in pursuance of the Danish Land Distribution Act (Jordfordelingsloven), Fødevareministeriet (2010). The Danish Land Distribution Act offers the affected farmers the opportunity to establish a land distribution solution that takes into account all involved parties to the extent possible. For instance, farmers looking to produce organic crops can trade or buy their way into the groundwater protection zones, where they are compensated by the water works for avoiding field spraying. A land distribution process requires a certain volume of participating farms to be successful. Land distribution processes are organised around a planner who may be a land surveyor, an agricultural consultant, or similar. The planner does the negotiations with each landowner. Thus, there is no direct trade between the individual farmers and participation in the process is voluntary. Once all agreements have been made, the land distribution plan is presented to a Land Distribution Committee, which makes a binding decision on behalf of the farmers who have entered into the agreements.

6.5.5   Urban efforts – information campaigns

Reports on pesticide findings have demonstrated that pesticides are found just as frequently in abstraction wells in urban areas as in abstraction wells located in agricultural areas, even though pesticide sales to private individuals only comprise 1% of the total sales in Denmark. Pesticide pollution constitutes a considerable problem despite the low share sold to private individuals. This is because private house owners in urban areas probably tend to use the pesticides on gravel-covered and tiled areas where the microbiological turnover is low. Furthermore, due to insufficient knowledge of appropriate application rates private individuals frequently overuse when applying pesticides.

A total of 16% of the groundwater-forming catchment areas of Aarhus Water's well fields are located in urban areas and targeted information campaigns are also implemented in these areas in line with the rural area campaigns. The Danish environmental legislation only provides a legal basis for restrictions in the form of pesticide-free production injunctions against commercial (professional) activities, hence information campaigns are used to change the behaviour of private garden owners. The exception to this rule are well field protection zones where the legislation also provides an opportunity to impose restrictions on private garden owners. The public water service companies in bigger cities are advocating for an amendment of the legal framework, providing the legal basis for restrictions on private garden owners in groundwater protection zones.

In urban campaigns, working groups are formed consisting of people from the local water service provider and the local citizens' council. The activities are typically of an informative nature, including: issue of information folders, publication of articles in local magazines and papers, and production of educational material for local schools and childcare centres. Questionnaire surveys performed in connection with urban campaigns have demonstrated that many private garden owners are unaware that they live on an aquifer and that their use of pesticides puts their drinking water at risk.

6.5.6   Efforts made on publicly owned areas

In 1997, the City Council decided to stop using pesticides on areas owned by the municipality located within high priority areas. The municipality owns roads, some municipal property, and about 2000 ha land.

The farmland that the municipality owns is leased on lease agreements, stipulating that no pesticides may be used on the land. In exceptional cases leaseholders may be allowed an exemption when spraying for common wild oat is needed. The municipality assesses that the lease revenues generated from the larger, continuous areas are EUR150 lower per ha because the land may be used only for pesticide-free production. The revenue reduction per ha for slightly smaller, cohesive areas is assessed at EUR65–130 and on small areas of less than 1 ha it is negligible.

The decision made by the Aarhus Municipality to end the use of pesticides is in line with the 'Agreement on the continual phase-out of the use of plant protection products on public areas', which was signed by the Danish Ministry of the Environment, the Danish Regions, and all Municipalities in Denmark in 2007 and herewith replacing a similar agreement from 1998.

6.5.7 Advice on field spraying equipment and avoiding point sources

Farmers who do their own field spraying are offered a review of spraying routines and equipment by Aarhus Water in collaboration with consultants from the local agricultural advisory associations. In addition, farmers were previously offered financial aid and advice on how to establish washing sites where the water is collected in safe tanks. At the washing site the farmers can fill their pesticide field sprayer and thereby avoid risk of point source pollution. Meanwhile, new legislation has made these efforts obsolete (see Section 6.5.8 Inspection of washing sites and field spraying equipment).

6.5.8 Inspection of washing sites and field spraying equipment

In Denmark, authorities and the agricultural sector have been actively involved in exposing the risks of pesticide waste and pollution in connection with crop spraying and in developing routines and equipment that minimise the risk that spraying work causes waste and pollution. This work has, among others, formed the basis for legislation that came into force in 2010 and which considerably strengthened the work to minimise the risk of point source pollution due to crop spraying (Danish Ministry of the Environment, 2015). A pivotal part of the legislation is that anyone performing field spraying in an occupational capacity shall hold a valid spraying certificate. To qualify farmers need to complete a 74-hour course, comprising legislation, spraying techniques, knowledge about new pesticides, decision support systems, information on pesticide labels, and integrated plant protection. The course concludes with an exam and sprayers attend a mandatory follow-up course once every 4 years.

Furthermore, the legislation established provisions on where and how the farmer may fill, empty, and wash the sprayer, along with other requirements, concerning sprayer equipment. For instance, the sprayer may be filled or washed only at an approved concrete site equipped with drainage, collecting the water in a sealable container or alternatively in always different locations in the field where the bioactive soil helps degrading the pesticides. Furthermore, there are requirements for the mixture refill equipment on the sprayer. Additionally, the sprayer shall be equipped with a rinse water tank and jets to wash out any remaining spraying product to ensure that when the spraying task has been completed any remaining spraying product can be watered down and sprayed on the treated area while the sprayer is moving. All spraying equipment for professional use shall be inspected at an approved inspection company (Danish Ministry of the Environment, 2014).

The inspection of the handling of pesticides, including the establishment of washing sites, is handled by the Ministry of Food, Agriculture, and Fisheries and is limited to approx. some 600 inspections annually in Denmark. On average, this means that Aarhus Municipality may expect about six annual inspections. The inspection is thus very limited and its effect as a preventive measure to avoid pesticide point source pollution may be considered largely absent. If Aarhus Municipality learns about pesticide point sources, e.g. via analyses of water samples, the municipality may issue an injunction for investigation and mitigation measures under the Soil Contamination Act.

6.5.9   Funding of groundwater protection

Groundwater protection work is complicated and expensive. The 25 or so water service providers in the municipality have joined forces to collaborate on funding and implementing groundwater protection. Collaboration of the water works in Aarhus is crucial for several reasons, the most important being: A. The groundwater catchment areas of the various well fields overlap. B. The Municipality of Aarhus has stipulated that groundwater protection zones classified as high priority area for drinking water shall be protected. This means that areas that are not related to present abstraction but are reserved for future abstraction must also be protected through agreements presented by the water service providers.

The current costs of groundwater protection are predominantly due to the fact that farmers are entitled to full compensation for any loss associated with their conversion to pesticide-free production. The cost corresponds to EUR0.07/m^3 of produced drinking water. It is expected that the overwhelming majority of the efforts will be completed within the next 15–20 years. By comparison, the total costs associated with production of drinking water in 2014 were EUR1.5/m^3, so the cost of groundwater protection is less than 5% of the cost of production.

The water service providers assess that protection of the groundwater is the most cost effective way to maintain the possibility of using uncontaminated groundwater for drinking water without any form of water treatment. Without groundwater protection it will likely not be possible to abstract a sufficient amount of uncontaminated groundwater within the municipality, hence water production costs increase. Current groundwater abstraction coupled with effective groundwater protection is a much more cost effective way to produce drinking water than other drinking water production method, like desalination of seawater or the use of treated freshwater from lakes (personal communication).

6.5.10   Differentiation of pesticide sources (point verses diffuse)

The occurrence of pesticides in groundwater may either have originated from diffuse or from point sources, which to a considerable extent defines the relevant action measures and also determines which authority is to take action. For administrative purposes a distinction is made between point sources (which come under the Soil Contamination Act) and diffuse sources from productive/urban areas where there are legitimate uses of pesticides. A considerable need has existed to develop/identify methods that enable

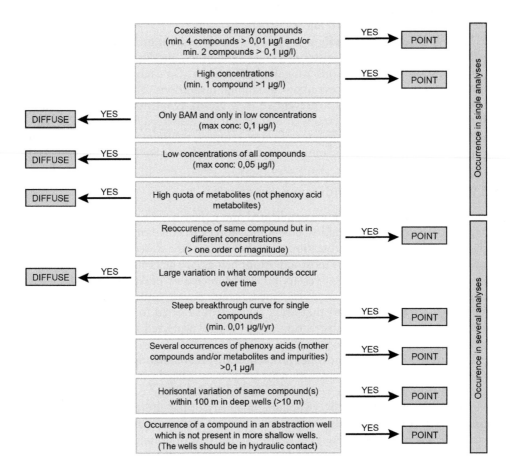

Figure 6.4 Decision support system for the assessment of the cause of pesticide occurrence in groundwater. The system assesses whether it is more likely that the contamination originated from a point source or a diffuse source. The assessment is made on the basis of the pesticide content of one or more water samples (Tuxen *et al.*, 2013).

the authorities to decide if pesticide occurrence in groundwater (e.g. at a monitoring point or an abstraction well) is due to a diffuse or a point source.

Some indicators may point to both types of sources. It is therefore important to assess the findings by deploying several indicators, as a single indicator will frequently be insufficient. Point sources are characterised by high concentrations in limited areas and diffuse sources by low contractions in large areas. The various point and diffuse source indicators are presented in Figure 6.4 as a flow diagram. The flow diagram distinguishes between indicators relating to single-analysis and multiple-analysis (e.g. time series or detection in several wells/filters. The indicators are presented only as 'Yes' tests. This implies that a 'No' does not imply a statement to the contrary (Tuxen *et al.*, 2013).

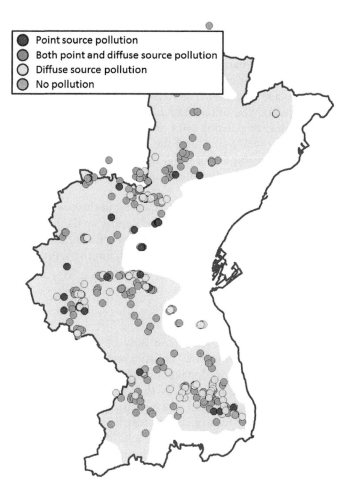

Figure 6.5 Map of Aarhus Municipality, presenting an assessment of where pesticide pollution originating from point sources or diffuse sources are found. Wells with no findings of pesticides are also denoted by blue dots. Light blue area are high priority areas for drinking water.

The indicators were used in the Municipality of Aarhus where 137 wells/ abstraction points with pesticide detection were investigated, using the screening tool. The screening indicated (Figure 6.5) that diffuse pollution caused pesticide contamination in 91 wells/abstraction points, corresponding to 66% of the pesticide contaminations in the municipality. Point sources caused pesticide contamination in 38 wells/abstraction points, corresponding to 28% of the pesticide contaminations in the municipality. The pesticide findings in the remaining eight wells could be attributed to both point and diffuse source contamination (6%). The screening results demonstrate a need to introduce measures against diffuse sources and therefore against field spraying in the groundwater protection zones. The screening results also identify a need for preventing point source pollution as well as taking measures to find and manage historical point sources.

6.5.11   Historic point source pollution

The mapping and remediation of historic point source pollution is the responsibility of the five Danish regions. The regions' efforts have been focused on anthropogenic substances like chlorinated solvents or oil spill products, but recently they have started taking pesticides into consideration. This is seen as an important step for the Aarhus area where 28% of pesticide findings in the wells are related to point source pollution. In the Aarhus Municipality, the authorities have investigated many point sources, mostly former landfill sites. Two major cases of point source pollution with pesticides constitute an acute risk to well fields, and the Region has performed screenings of possible sources in one of the important groundwater catchment areas. Mapping and handling of historic point source pollutions is an area that needs development of new strategies and techniques.

Aarhus Water is performing remediation pumping in three well fields in an effort to stop pesticide point pollution impacting water supply wells. In one of these cases the point source is well described and originates from spills on a nearby farm. In one of the other cases an entire well field has been polluted, but the source has not yet been established despite great efforts.

6.6   CONCLUSIONS

Groundwater in the Aarhus Municipality can only be protected against pesticide/nutrient contamination by a long-term, holistic, effort, involving water service providers, management agencies, and stakeholders. This has necessitated development of new geophysical methods to map clay thickness to assess the risk to aquifers from surface activities (both agricultural and urban), extensive monitoring of both groundwater level and quality, extensive stakeholder liaison and communication, and the development of multiple pieces of legislation, including multiple amendments. This has taken decades at a considerable cost, but without this level of endeavour optimal groundwater management would not be possible to protect Denmark, and in particular Aarhus's groundwater resources.

During the course of the 1990s it was acknowledged that approx. one third of the general abstraction wells in Aarhus Municipality were contaminated with pesticides. Aarhus University, Aarhus County, and Aarhus Water collaborated extensively during this period on technology development and mapping. In 1998, a new act was passed in Denmark on mapping of vulnerable areas, identification of groundwater protection zones, and action plans for such zones. The experiences from the Aarhus area were employed in the preparation of the act. The methods to map groundwater vulnerability initially focused on nitrate, which was a known pollution issue as early as the 1980s, both in Denmark and in the EU.

Throughout the process there has been an understanding that pesticide vulnerability is a more complex issue than nitrate vulnerability. However, as they both commonly occur together, the presence of nitrate in groundwater may be a good indicator of pesticide risk. The mapping activities have shown that in the Aarhus area pesticide vulnerability and nitrate vulnerability coincide extensively.

It was decided to designate the nitrate vulnerable areas as groundwater protection zones with respect to nitrate as well as pesticides. There is only a limited need for efforts targeting nitrate, as the drinking water quality threshold criteria are only exceeded in limited areas. The need for efforts targeting pesticides is more widespread as drinking water quality thresholds are exceeded in many areas. At the political level there is an ambition that all vulnerable areas should be protected by imposing a ban on the use of pesticides in these areas.

Such a ban may be implemented under the provisions of the 1998 Act. Initially, the water works are to offer voluntary agreements, but if stakeholders fail to enter into a voluntary agreement the municipality can issue a pesticide ban. Bans can be issued in well protection zones as well as groundwater protection zones.

Since 1998, the water service providers have been offering voluntary agreements and providing advice on conversion into pesticide free farming. The efforts have been prolonged and persistent and many agreements have been made. Nevertheless, the voluntary agreements only cover a fraction of the vulnerable areas. The local authorities have decided to exploit the facilities of the legal framework to introduce binding requirements on pesticide-free production and in 2013 passed the first ever action plan, which makes use of injunctions. Based on the experience in Aarhus water quality protection through voluntary agreement is unlikely to succeed. A significant percentage of land holders in this jurisdiction only changed behaviour as a result of enforced pesticide bans and injunctions.

The success of the water quality protection process in Aarhus has resulted from the interaction of a number of factors. The strength of Aarhus's professional hydrogeological community, including the university, the water service providers, and authorities to develop the technology and methods, which facilitated vulnerability mapping in the 1990s. Another essential component is the political will, present both at the local and national level to ensure drinking water protection by providing programme stability with sufficient long term resources. This has led to national legislation, including mandatory instruments and passing of local action plans for the protection of the drinking water against pesticides.

REFERENCES

Aarhus Kommune (2013) *Indsatsplan Beder* [Online] Available from: http://www.aarhus.dk/~/media/Dokumenter/Teknik-og-Miljoe/Natur-og-Miljoe/Vand/Grundvand/Beder/Indsatsplan-Beder-VEDTAGET-17-april-2013.pdf

DANVA (2000) Dyrkningsaftaler og kompensationer, [Online] Available from: http://www.energibesparelser-vand.dk/Div.%20aftaler-1979.aspx

EU (2006) Europaparlamentets og Rådets Direktiv 2006/118/EF af 12. december 2006 *om beskyttelse af grundvandet mod forurening og forringelser.* EFT L 327 af 22.12.2000. pp. 1–72 og 10 bilag. (Grundvandsdirektivet)

Fødevareministeriet (2010) *Bekendtgørelse af lov om jordfordeling og offentligt køb og salg af fast ejendom til jordbrugsmæssige formål m.m. (jordfordelingsloven)* [Online] Available from: https://www. retsinformation.dk/Forms/R0710.aspx?id=133746

Fødevareministeriet (2010) Land distribution Act: *Bekendtgørelse af lov om jordfordeling og offentligt køb og salg af fast ejendom til jordbrugsmæssige formål m.m. (jordfordelingsloven).* LBK nr 1275 af 02/11/2010. [Online] Available from: https://www.retsinformation.dk/forms/R0710.aspx?id=133746

GEUS (2015a) page for groundwater protection. [Online] Available from: http://www.geus.dk/DK/popular-geology/edu/viden_om/grundvand/Sider/gv03-dk.aspx

GEUS (2015b) page for data and maps. [Online] Available from: http://www.geus.dk/UK/data-maps/Pages/default.aspx

GEUS (2015c) *The National Groundwater Monitoring Programme.* [Online] Available from: http://www.geus.dk/UK/water-soil/monitoring/groundwater-monitoring/Sider/default. aspx [Accessed 14th January 2015].

Hansen B., Thorling L., Dalgaard T., & Erlandsen M. (2011) *Trend Reversal of Nitrate in Danish Groundwater – a Reflection of Agricultural Practices and Nitrogen Surpluses since 1950.* Environmental Science and Technology, 45 (1), 228–234.

Homepage Geological Survey of Denmark and Greenland. [Online] Available from: http://www.geus.dk/UK/data-maps/Pages/default.aspx

Miljøministeriet (2010) *Bekendtgørelse af lov om miljøbeskyttelse* [Online] Available from: https://www.retsinformation.dk/forms/r0710.aspx?id=132218

Miljøministeriet (2013) *Bekendtgørelse af lov om vandforsyning* [Online] Available from: https://www.retsinformation.dk/forms/r0710.aspx?id=145854

Miljøministeriet (2014a) *Bekendtgørelse om påfyldning og vask m.v. af sprøjter til udbringning af plantebeskyttelsesmidler* [Online] Available from: https://www.retsinformation.dk/Forms/R0710.aspx?id=165124&exp=1

Miljøministeriet (2014b) *Bekendtgørelse om syn af sprøjteudstyr. BEK nr 1430 af 16/12/2014* [Online] Available from: https://www.retsinformation.dk/forms/R0710.aspx?id=166660

Miljøministeriet (2015) *Bekendtgørelse om påfyldning og vask m.v. af sprøjter til udbringning af plantebeskyttelsesmidler. BEK nr 1598 af 18/12/2014 (Vaskepladsbekendtgørelsen).* Miljøministeriet [Online] Available from: https://www.retsinformation.dk/forms/R0710. aspx?id=165124

Municipality of Aarhus (2010) *Vand Vision 2100* [Online] Available from: https://www.aarhus. dk/~/media/Dokumenter/Teknik-og-Miljoe/Natur-og-Miljoe/Planer-og-projekter/Vandvision 2100.ashx

Nygaard, E. (red) (2004) *Koncept for Udpegning af Pesticidfølsomme Arealer, KUPA. Særligt pesticidfølsomme sandområder: Forudsætninger og metoder for zonering.* GEUS. [Online] Available from: http://kupa.dk/xpdf/KUPA_sand_slutrapport.pdf

Thomsen, R., & Søndergaard, V. (2007) *Dense hydrogeological mapping as a basis for establishing groundwater vulnerability maps in Denmark.* Groundwater Vulnerability Assessment and Mapping – Selected papers from the Groundwater Vulnerability Assessment and Mapping International Conference, Ustron, Poland, 2004. International Association of Hydrogeologists, Volume 11, Chapter 3: 33–43.

Thomsen, R., Søndergaard, V.H., & Sørensen, K.I. (2004) *Hydrogeological mapping as a basis for establishing site-specific groundwater protection zones in Denmark.* Hydrologeology Journal 12, 500–562.

Tuxen N., Roost, S., Kofoed, J.L.L., Aisopou, A., Binning, P.J., Chambon J., Bjerg, P.L., Thorling, L., Brüsch, W., & Esbensen, K. (2013) *Skelnen mellem pesticidkilder.* Danish EPA (Miljøstyrelsen). Miljøprojekt nr. 1502 [Online] Available from: http://mst.dk/ service/publikationer/publikationsarkiv/2013/okt/skelnen-mellem-pesticidkilder/

Thomsen, R., Sondergaard, V. (Tech. Ed.) & Klee, P. (Ed.) (2013) *Greater water security with groundwater – Groundwater mapping and sustainable groundwater management. The Rethink Water network and Danish Water Forum white papers,* Copenhagen. [Online] Available from: www.rethinkwater.dk/whitepapers/ or: https://stateofgreen.com/en at: https://stateofgreen.com/files/download/1152

Chapter 7

Models, decision-making and science

J. Doherty[1] & R. Vogwill[2]

[1] Professor, Flinders University, National Centre for Groundwater Research and Training, Adelaide, South Australia
Director, Watermark Numerical Computing, Brisbane, Queensland, Australia
[2] Associate Professor, School of Earth and Environment, The University of Western Australia, Crawley, Australia

ABSTRACT

Management of groundwater systems relies heavily on groundwater models. They are often commissioned by one party and then used by another. Assurance of a model's quality often rests on compliance with guidelines or standards. This chapter argues that integrity of model-based decision-support requires more than this. It argues that the decision-making process requires nothing less of modelling than that it implements the scientific method. This requires recognition of the stochastic nature of expert knowledge on the one hand, and the limitations of history matching in refining that knowledge on the other hand. Model predictions must therefore be viewed as probabilistic. As such they can form a basis for risk assessment, this being a vital component of any decision-making process. To fulfil this role models must be used in partnership with equally sophisticated stochastic and inversion software. However, regardless of the level of modelling sophistication, assessment of risk can only ever be subjective. In making the innumerable subjective decisions that the modelling process demands, a modeller's reference point must necessarily be avoidance of failure of the modelling exercise. This happens if the risk of occurrence of an unwanted event is assessed to be lower than it actually is.

7.1 INTRODUCTION

This chapter examines the role that environmental models in general and groundwater models in particular should play in the environmental decision-making process. It is the author's contention that such an examination is urgently required as current practice, often reinforced by official and unofficial modelling standards and guidelines, is not serving that process well. The potential for confusion among those who develop models and those who must interpret model predictions in the context of their decision-making imperatives is high. So too is the potential for misuse of those predictions.

The construction of a groundwater model, which is built to support and illuminate environmental decision-making requires the making of many decisions itself. These decisions pertain, among other things, to: numerical and conceptual complexity; the mathematical devices that are used to represent hydraulic and geochemical property heterogeneity throughout its domain; specification of hydraulic and geochemical boundary conditions; the extent to which hydraulic properties should be adjusted through a history-matching process; and the level of fit with historical measurements of system state that should be sought through that process.

As for decisions made for any purpose, an intellectual framework must exist through which the benefits and costs associated with the many choices that comprise the model construction process can be assessed. (This assumes, of course, that decision makers wish to use models that are the fruits of such a logical construction process. This may not always be the case.) Despite the existence of multiple national and international standards and guidelines, such a framework has not been developed for groundwater modelling. Indeed, the high level of subjectivity that is necessarily associated with model construction suggests that it may be difficult to construct one. Nevertheless, this chapter attempts to provide a logical basis upon which to do so. The alternative is to continue along the present path, wherein the uncertainties associated with critical model predictions typically remain untested, appearances of 'model fidelity' count for more than substance, and model-based decision-making rests more on an illusion of predictive accuracy than on a quantitative assessment of risk.

Fundamental to an intellectual framework that supports any activity is a definition of failure. Without a definition of failure there are no criteria by which to judge success. Without a concept of failure there can be no means to convey to a client or stakeholder the bases for the innumerable choices that were made in building a model. Without it there can be no possibility of relevant review of one modeller's work by another. A definition of failure, as it applies to groundwater modelling undertaken in the decision-making framework, is also provided in the discussion that follows.

7.2 MANAGEMENT DECISIONS

The management and protection of groundwater systems, including man-made or natural systems that interact with them, such as mines or wetlands, requires that many decisions be made. They include: to pump or not to pump; ... where and how much to pump; where and how much to reinject. These are just a few of the options that are routinely considered by managers, investors, and regulators.

The fact that decisions must be made at all, implies that a groundwater system is at least partially under human, rather than natural management. This happens where there is a benefit in interfering with the natural operation of that system. In some instances this benefit is the direct use or clean-up of the groundwater resource. In other cases benefits flow from an activity (such as mining) that has interfered with the resource. In all cases an infrastructure cost is incurred, a cost which is more than compensated for by the typically financial benefits of interfering with the system. If this was not the case then presumably the desire to interfere with the groundwater system would not be present.

Benefits and costs form the building blocks of basic accounting. So too does the concept of risk. Freeze *et al.* (1990) define the role of the groundwater model in the decision-making process as one of risk assessment. Risk can be defined as the probability of something going wrong multiplied by its cost. Freeze *et al.* (1990) explain that an optimal decision is one that maximises an objective function φ, which is defined as something like this.

$$\varphi = B - C - R \qquad (7.1)$$

In equation (7.1) B are the benefits accrued through taking a particular course of action, while C are the direct costs of that action. R is the risk of something going wrong, i.e. the risk of an unwanted event. It is the role of the groundwater model to associate a probability or likelihood with an unwanted event, particularly events that constitute environmental impact (which are frequent targets of risk assessment). Society, and/or the proponent of a project, assign a cost to that event. If the cost is high, then the unwanted event, or 'bad thing', the focus of risk evaluation must be seen as having a low probability of occurrence if a project or particular management strategy is to proceed. If the cost is low, then a project may proceed with a lower level of guarantee that the bad thing will not happen. Note that the assignment of cost to an event that is undesirable from a social or environmental perspective is not always straightforward. Wallace (2007) proposed a framework to evaluate environmental and social costs so that trade-offs with economic benefit could be evaluated. The evaluation of environmental and social costs can be a contentious matter (Costanza & Kubiszewski, 2012). However, it cannot be avoided.

The notion of ascribing a probability of occurrence to a bad thing can be extended to complex environmental management processes, such as that often described as 'adaptive management'. It may not be possible to say upfront that a project or course of action that accrues substantial economic benefits may not also precipitate the occurrence of a bad thing. However, this may not be sufficient grounds for a regulator or an investor to scrap the project. A groundwater (and/or other) monitoring network may then be built to provide early warning of an unwanted occurrence at the same time as it provides a better understanding of a complex system. It may be decided in advance that if observations of groundwater level or quality within a monitoring well reach a certain threshold, then the project proponents must undertake remedial action of a certain type (including possibly cessation of the project). The task of a groundwater model then becomes that of establishing that the agreed-upon remedial action precipitated by the breaching of the threshold will, at a certain level of confidence, prevent the occurrence of the bad thing. This allows all parties to acknowledge that both the criteria and remedial measure are suitable, despite shortcomings in current understandings of cause-effect linkages that constitute an environmental system.

In its most simple terms, then it is the task of groundwater modelling to establish whether a bad thing can be avoided if a certain course of management action is undertaken. This requires that consideration of a course of management action be accompanied by the proposal of a hypothesis that this action will be accompanied by the occurrence of an unwanted system state or event, i.e. by the occurrence of a bad thing. The groundwater modelling process must then attempt to provide the basis for rejection of that hypothesis at a certain level of confidence. In doing this a groundwater modeller is doing nothing more than implementing the scientific method. Their model then becomes a scientific instrument.

On what grounds then can the hypothesis that a bad thing will happen be rejected and how is a groundwater model equipped to do this? The hypothesis that a bad thing will happen can be dismissed as improbable if its occurrence is shown to be incompatible with what is known about the properties of a system and/or with what is known about the historical behaviour of that system. A groundwater model is unique in providing the basis for such an inquiry. Furthermore, as will be discussed below, its ability to perform this function can be optimised if the principles of its construction are the same as that of any other scientific instrument. Hence, it must be tuned to this prediction-specific task, possibly at a cost to its ability to perform other tasks.

It therefore follows that the role of a groundwater model is to explore possibilities. Its *modus operandi* must embrace the fact that predictions of future environmental behaviour cannot be made with certainty. It must explore predictive possibilities rather than 'make predictions', with exploration of these possibilities resting on its unique ability to provide receptacles for two types of information, namely expert knowledge on the one hand and the information contained in historical measurements of system state on the other hand.

7.2.1 Failure

If the purpose of a model is to test the hypothesis that a bad thing will accompany the adoption of a certain management strategy, then the definition of failure of a modelling exercise becomes obvious. A modelling exercise fails when a false rejection occurs. If the outcome of a modelling exercise is the conclusion that a bad thing will not happen, or that its occurrence is very unlikely, then that exercise has failed if the likelihood of the bad thing's occurrence is higher than concluded. To put it in simple terms, if modelling rejects the hypothesis that a bad thing can happen and then the bad thing actually happens, this is failure.

This definition of failure is aligned with the definition of a 'type II statistical error'—the false rejection of a hypothesis. Given the nature of the hypothesis that requires testing in the decision-making context, the repercussions of modelling failure for that process are obvious.

With the above outline of the place of modelling in decision-making and with this definition of failure, model-based implementation of the scientific method becomes perfectly aligned with model-based decision-support. Nothing should be asked of a model beyond that, which the scientific method can promise. Despite the misapprehensions of many of those who pay for models (misapprehensions that are rarely corrected by those who build models for payment) a model cannot tell us what will happen in the future as too little is known of too many aspects of the system which it simulates, a matter that is discussed in greater detail below. However, if constructed as a scientific instrument, it may possess the ability to tell us what will *not* happen in the future. The ability to rank proposed management strategies in terms of their ability to guarantee avoidance of a bad thing (at costs that are also assessed as part of the decision-making process) is exactly what groundwater management in particular and environmental management in general requires.

7.2.2 Hypothesis-testing and uncertainty analysis

It was stated above that a model's ability to test and possibly reject hypotheses of management interest rests on its ability to provide receptacles for two types of information, namely expert knowledge and the information that is resident in historical measurements of system state. Let us take a moment to examine the nature of these two types of information.

Expert knowledge, especially when applied to subsurface hydraulic properties and processes, is necessarily stochastic in nature. A geologist does not know what is under the ground except at the locations of a small number of wells. Spatial interpolation of intersected lithologies and the assignment and interpolation of hydraulic properties between points of lithology intersection is a highly uncertain business. Nevertheless, there are bounds on this uncertainty. A geologist will quickly inform a modeller of the error of their ways if a model encapsulates stratigraphic boundaries, which violate geological principles or if hydraulic properties are assigned to model cells or elements which contravene those normally associated with prevailing rock types. Hence, in harmony with its stochastic nature, expert knowledge can provide assurances of what is *not* down there. At the same time as it cannot provide assurances of what *is* down there.

Geostatistical software and concepts can be used to express geological expert knowledge, including the innately stochastic nature of that knowledge. In early geostatistical software, the statistical expression of spatial hydraulic property variability was restricted to multi-Gaussian fields, e.g. Deutsch & Journel (1998). Today, geostatistical simulators are much more complex. Realisations of geology and hydrogeological properties are often categorical in nature, with each such realisation exhibiting a different disposition of discrete stratigraphic units, depositional features (such as alluvial channels and crevasse splays), or of structural features, such as faults and fracture networks. Discrete stratigraphic units may be populated, in turn, by realisations of hydraulic properties, which are multi-Gaussian in nature or which possess some other statistical basis. At the same time multiple point geostatistics has supplanted two-point statistics as descriptors of spatial variability, as the former can better represent the types of heterogeneity that characterise geological media more realistically than the latter. For example see Remy *et al.* (2009).

Expert knowledge can be expressed in a groundwater model by running the model many times. On each occasion that the model is run it is populated by a different geostatistical realisation of geological features and structures, together with a different realisation of hydraulic properties attributed to those features and structures. Ideally, the probability distribution of a model prediction of management interest can then be empirically determined through collecting predictions made, using all of these realisations and constructing a histogram for that prediction. Unfortunately, a prohibitively large number of model runs may be required to properly characterise probabilities associated with the extremes of the resulting probability distribution.

As well as being run under predictive conditions, a model can also be run under historical conditions, using these same geostatistical expressions of expert knowledge. In most modelling contexts, few if any model outputs that correspond to historical measurements of system state would actually reproduce those historical measurements. Suppose, however, that the ability to undertake an effectively unlimited number of model runs is available to a modeller. Then geostatistical realisations which do not allow a model to replicate measurements of historical system behaviour could be rejected, leaving the modeller with only those that do. This remaining set of 'behavioural' (Beven, 2009) realisations could be used to explore the uncertainties of model predictions. These behavioural realisations express expert knowledge (including its stochastic nature), but also respect the constraints imposed by the necessity for the model to reproduce the past. Hence, they encapsulate the information contained within historical measurements of system state.

This assimilation of the two forms of knowledge that a model requires to test hypotheses pertaining to future system behaviour is expressed by Bayes equation. In its simplest form it can be written as follows. In this equation the term $P(x)$ can be interpreted as 'the probability of x'.

$$P(\mathbf{k}|\mathbf{h}) \propto P(\mathbf{h}|\mathbf{k})P(\mathbf{k}) \tag{7.2}$$

The last term on the right of equation (7.2) refers to the so-called 'prior probability distribution' of model parameters, processes, boundary conditions, and all other aspects of model design. For the sake of simplicity these are refered to as 'parameters' in the discussion that follows and described using the vector $\mathbf{k}$. The term 'parameters' is thus used in the broadest sense. This rightmost term of Bayes equation encapsulates expert knowledge.

The first term on the right of Bayes equation is the so-called 'likelihood function'. Strictly speaking, it describes the probability of observing the measurement dataset (described by the vector $\mathbf{h}$), given a particular set of parameters $\mathbf{k}$. The better the fit between model outcomes and historical measurements is, the higher is the value of this term. It can thereby act as a filter on the prior parameter distribution. However, this fit can never be perfect because observations are accompanied by measurement noise and because model outputs reflect model imperfections. Furthermore, even if a perfect fit between model outputs and a corresponding measurement dataset could be obtained on the basis of a particular $\mathbf{k}$, in the groundwater modelling context this same fit could also be obtained using an infinite number of other $\mathbf{k}$s whose vector-differences with the original $\mathbf{k}$ occupy a high dimensional, so-called 'null subspace' of parameter space. See Moore & Doherty (2006) for details.

The term on the left of Bayes equation is the posterior parameter distribution. Literally it says 'the probability of parameters $\mathbf{k}$ conditioned by the observation dataset $\mathbf{h}$'. This term expresses the fact that model parameters and the predictions that depend on them can only be probabilistic in nature. Hence, every prediction made by a model is accompanied by at least some uncertainty. This uncertainty reflects the stochastic nature of expert knowledge on the one hand and the ability (or lack of ability) of historical observations of system behaviour to refine that knowledge on the other hand.

Different predictions made by the same model will be accompanied by different levels of uncertainty. Moore & Doherty (2005) show that the constraints on expert-knowledge-based parameter variability imposed by an historical observation dataset can reduce the uncertainties of some predictions to relatively small levels. At the same time, however, they can leave the uncertainties of other predictions relatively undiminished from their prehistory-matching state where the only constraints on those predictions are those imposed on parameters by expert knowledge. In general, those predictions which are sensitive to areal averages of parameter values and/or those which most resemble the measurement dataset in type and in nature of system stresses, will be those which are most informed by the measurement dataset. In contrast, predictions that depend on parameter detail and predictions, which reflect a different set of system stresses from those which were operative at times during which measurements were made, will tend to be those whose uncertainties are least reduced through history-matching (if at all).

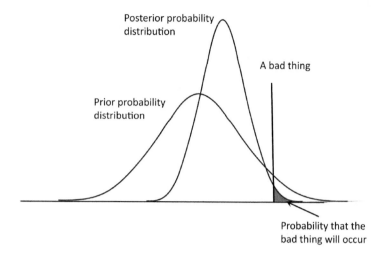

Figure 7.1 Defining the probability of occurrence of a bad thing.

Once an uncertainty distribution has been associated with a model prediction, the chances of its occurrence can be evaluated. Strictly speaking, the hypothesis of its occurrence can only be rejected at a certain probability level. This is shown schematically in Figure 7.1. The black bar marks an unwanted predictive occurrence. The area under the posterior probability distribution (as under all probability distributions) equals 1. The area under the posterior probability distribution to the right of the black bar is the probability that the value of a future prediction will exceed an unwanted threshold so that the bad thing occurs.

There are more efficient ways to sample the posterior predictive probability distribution than the rejection sampling methodology described above. The Markov chain Monte Carlo (MCMC) method is designed specifically for this purpose. Sadly, despite huge efficiency gains over rejection sampling (cf. Laloy & Vrugt, 2012), the computational cost of MCMC is normally very high. This is especially the case where predictions of interest are sensitive to parameterisation detail, so that the number of parameters that must feature in the analysis is therefore necessarily large. Furthermore, despite some significant advances over the last few years, efficient use of MCMC with geostatistically-based parameter fields is a procedure that is still in its infancy.

More direct methods of hypothesis testing are also available. For example parameter integrity and model-to-measurement compatibility can be assessed while adjusting model parameters in a history-matching process, in which a historical measurement dataset is supplemented with a hypothesised observation of an unwanted future event, see for example Moore *et al.* (2010) and Doherty (2014). Unfortunately this methodology is also computationally expensive and encapsulates expert knowledge in an abstract manner that may erode, to some extent, its information content.

7.2.3 Model calibration

Rightly or wrongly, the ubiquitous 'calibrated groundwater model' is pivotal to groundwater management worldwide. This seems to contradict ideas that have been presented so far in this chapter, as the word 'calibration' implies parameter uniqueness, and conveys a sense of (at least attempted) predictive certainty. This raises some important questions. If expert knowledge is necessarily stochastic, and if history-matching does not eliminate, and may even do little to constrain, the stochastic nature of model predictions, what role can the 'calibrated model' legitimately play in groundwater management? Why pursue a quest for uniqueness – a quest which is fundamental to the quest for a calibrated model – when testing hypotheses requires that uncertainty be embraced and that all predictive possibilities be explored?

Sadly, much of the faith that is placed in 'the calibrated model' has little scientific basis. It is an outcome of habit. Modellers are told by managers 'we need a single number' and that 'things will only get confused if we barrage stakeholders with fuzzy things like probability distributions'. Managers may indulge modellers with further elaboration of their requirements of the modelling process with statements such as: 'If stakeholders can see that the calibrated model replicates the past, then they will believe us when we say that it can predict the future. Let's keep it all simple and definitive, otherwise there will be no possibility of reaching consensus.' In some cases there are more insidious reasons for embracing uniqueness. Of the many parameter fields that may endow a model with the ability to replicate past system behaviour, a single one may be chosen for the making of model predictions because the values of these predictions are favourable to one of the parties in a dispute. Challenging the culture of the calibrated model is difficult or impossible when those who set model specifications are also those who pay for models to be built. This culture of 'the calibrated model' is so deeply ingrained and the repercussions of challenging this culture so dire, that few modellers have any desire to question it.

Nevertheless, model calibration is not without its benefits. This is because strict implementation of Bayes equation requires that a modeller work only with parameter probability distributions. This is numerically difficult, with the level of difficulty increasing rapidly with rising model run times that inevitably accompany increased model complexity and enhanced parameterisation sophistication. In contrast, even in a highly parameterised context, solution of an inverse problem for the attainment of a single parameter field is often numerically feasible. Given these practicalities the quest for uniqueness should then become the quest for a parameter field that is of minimum error variance. Such a parameter field allows a model to make predictions, which are also of minimum error variance. Post-calibration predictive error analysis, centred on parameters and predictions of minimised error variance, can then legitimately replace posterior predictive uncertainty analysis in exploration of the range of predictive possibilities that are compatible with all that is known of a system.

Using a combination of Tikhonov and subspace regularisation methodologies it is indeed possible in most decision-making contexts to compute a unique parameter field that can lay claim to a status of minimum error variance. For details, see Aster *et al.* (2005), Moore & Doherty (2005), and Tonkin & Doherty (2005). If regularisation is properly implemented, predictions made using this parameter field are thereby expected to lie somewhere near the centre of the posterior predictive probability distribution. It is important to note, however, that the claim of minimised predictive error variance is not a claim that the potential for predictive error is small. The claim is only that predictions made by a thus-calibrated model are unbiased and are therefore roughly centred with respect to the posterior predictive probability distribution. The potential for predictive error can then be analysed relatively efficiently, using subspace methods such as the null space Monte Carlo method described by Tonkin & Doherty (2009) and Doherty (2014). As is demonstrated by Keating *et al.* (2010), sampling the post-calibration predictive error distribution provides a reasonable approximation to sampling the posterior predictive uncertainty distribution, but is numerically far cheaper.

What is important is that 'the calibrated model' should not be pursued as an end in itself. Rather, the model calibration process should be considered as the first step in a two-step process of inversion (i.e. calibration), followed by predictive error analysis. Solution of the inverse problem should be undertaken using context-optimised regularisation methodologies that promise a calibrated parameter field that approaches that of minimum error variance and can thus underpin the second step of the modelling process. Predictive error analysis replaces predictive uncertainty analysis in implementing this second step. In the decision-making context, the metric for failure of this entire process is no different from that, which would prevail if the entire analysis were to be undertaken, using purely Bayesian methods, namely the occurrence of a type II statistical error. In general, the potential for this type of error can be minimised if all parameterisation detail to which a prediction is sensitive is included in the above two-step process, regardless of whether the first step in this process allows unique inference of such detail, e.g. Moore & Doherty (2005).

7.2.4 Model complexity

These concepts of modelling in decision-making can now be applied in addressing the important (and often divisive) issue of appropriate model complexity. Hopefully, with failure defined (i.e. incursion of a type II statistical error), a metric now exists against which complexity choices can be evaluated.

Appropriate complexity is both a prediction-specific and a decision-specific issue. If the model construction process is undertaken in such a way as to ensure that the uncertainty of a decision-critical prediction is overestimated, rather than underestimated, then a type II statistical error will be avoided. If, under a demonstrably conservative predictive uncertainty regime, an unwanted event can be awarded a low probability of occurrence, this may allow a desired project/management practice to proceed. Meanwhile the modelling process will have done its job.

Predictive uncertainty conservatism can be achieved in a number of ways. It may be achieved through forgoing a history-matching process, while at the same time ensuring that model outputs calculated on the basis of different realisations of expert knowledge do indeed span measurements of system state. Alternatively, expert knowledge may inform the use of parameters that are purposely chosen to promulgate pessimistic predictions through a 'worst case scenario' analysis. Another modelling option may be to employ a simplistic parameterisation scheme, with or without a parameter estimation phase, together with an additive 'predictive noise' term that can be demonstrated to encompass predictive errors incurred through model and parameterisation simplicity.

In all of these cases, if predictive uncertainty can be shown to be overestimated, and can be shown to account for any bias introduced to a decision-critical prediction through model simplification, then the grounds for rejection of a hypothesised bad thing are solid. The likelihood of model failure through incurring a type II statistical error is small. No greater model complexity is required.

It follows logically that the case for increased model complexity must rest on a desire to test and possibly reject hypotheses that cannot be rejected with a simpler model, given the need for predictive uncertainty conservatism in deployment of the latter model. However, it is important to recognise that the superior hypothesis-rejection ability of a more complex model does not necessarily reside in powers of 'greater predictive accuracy' that are often bestowed on complex models because they are construed to behave more like 'the real thing' or because their parameter fields are more picturesque. In contrast, it is the superior ability of the complex model to provide receptacles for either or both of expert knowledge and the information content of measurements of system state, that may promulgate a narrowing of the uncertainty intervals of management-critical model predictions. This, of course, assumes that these uncertainty intervals can be evaluated using the complex model.

This is where a quandary often arises. If a model is to provide a more comprehensive means of expressing expert knowledge then, given the stochastic nature of expert knowledge and geostatistical software as the means of its expression, the model must be used in partnership with such software. In doing this, the complex model must be run many times in order to explore predictive stochasticity as this arises from parameter stochasticity. Similarly, if a model is endowed with many parameters in order that its outputs may better match measurements of system state, so that it can internalise the information contained therein, then it must be run many times under the control of inversion software such a PEST (Doherty, 2014) or PEST++ (Welter et al., 2014). In either case, the potential that a complex model possesses for quantification of predictive uncertainty will not be realised if its run times are too long, as it will not be possible to undertake the number of model runs that stochastic analysis or the inversion process requires. When used with inversion software, good numerical behaviour is also required of a model so that derivatives of model outputs with respect to parameters can be computed with integrity. Computation of derivatives lies at the heart of highly-parameterised inversion.

Sadly, in so many instances, complex models offer little support to the decision making process because their run times are unusably long and their numerical behaviour is problematic. They can only be run a small number of times. Meanwhile their numerical delicateness dictates the use of parameter-specific solver settings. Given that the potential benefits of complexity can only be realised where a model can 'dance' with appropriate partner software, complexity that renders a model an unfit partner for any such software constitutes nothing more than a millstone around the model's neck. Such a model cannot be used to quantify and certainly not reduce the uncertainties of decision-critical predictions. At best such a model contributes little or nothing to the decision-making process; at worst the contributions that it makes to the decision-making process are misleading.

7.2.5 A fundamental choice

It has been argued above that a model's ability to contribute to environmental decision-making rests on its ability to provide repositories for two types of information, namely that resident in expert knowledge and that resident in historical measurements of system state. The former is achieved through a model's representation in a realistic manner of the components of an environmental system at a scale at which estimates of the properties of those components can be transferred to the model. The latter is achieved by adjusting a possibly large number of model parameters (usually with assistance from inversion software) in order to achieve a good fit between measurements of system state and corresponding model outputs.

It is an unfortunate fact that with present technology a groundwater model, which is capable of expressing expert knowledge is not so capable of replicating historical system behaviour. The converse is also true. This is because expression of geological expert knowledge generally requires a complex categorical parameterisation of subsurface hydraulic property heterogeneity, whereas the matching of field measurements is most readily achieved with continuously differentiable parameter fields, which suppress the sharper edges of subsurface heterogeneity. Notwithstanding some recent advances in algorithms through which categorical geostatistical realisations can be generated and adjusted, so that a model, which employs them is able to respect measurement constraints (see Zhou *et al.*, 2014 for a review), the technology is still not advanced enough for their regular use in everyday groundwater modelling practice. The right side of Bayes equation has only two terms. Sadly with present technology better representation of one of these terms comes at a cost of inferior representation of the other. In the groundwater modelling context, quantification and minimisation of predictive uncertainty is therefore a compromise and can never be exact.

A modeller is therefore forced to choose between two options for actual or implicit implementation of Bayes equation in lowering and quantifying the uncertainties of decision-critical predictions. As in the making of any choice, benefits must be weighed against costs. The cost of endowing a model with geologically realistic and necessarily stochastic parameter fields is an inability to constrain these fields very well through the history-matching process. While this may guarantee avoidance of a type II statistical error, it may limit the degree to which the uncertainties associated with decision-critical model predictions can be reduced. In contrast, the cost of employing smoother (though still necessarily heterogeneous and stochastic) parameter fields in place of more realistic 'picture perfect' geostatistically-based categorical parameter fields is that some guarantee must be found that use of the former in place of the latter does not artificially reduce the range of predictive possibilities or worse still incur an unknown degree of predictive bias as history-matching constraints are imposed. Either of these parameterisation strategies may incur a type II statistical error.

Ideally, the choice will rest on the relative amounts of information that can be accessed through both of these options. Where a model domain is relatively small and where much has been invested in site characterisation, the geostatistical option, possibly accompanied by weak history-matching, may prove more fruitful. This is often the preferred course of action in the reservoir modelling context. In contrast, where a model domain is large and where hydrostratigraphy is complex, a continuous parameter field will normally be adopted for each of a number of different hydrostrati-graphical units represented in a model domain, possibly layered and zoned to some extent in order to respect known vertical and horizontal facies variations and regional geological boundaries. Within each zone or layer a parameterisation device such as pilot points (Certes & de Marsily, 1991; Doherty, 2003) may be employed. Hydraulic properties assigned to these points can be readily adjusted as regularised inversion and then null space Monte Carlo methods are sequentially employed to firstly calibrate the model and then quantify the potential for posterior predictive error. This is the course of action that is often adopted in groundwater modelling practice.

But here the groundwater modeller is presented with yet another quandary. Through deployment of a simplistic, though dense parameterisation scheme based on devices such as pilot points, a model can be equipped with the means to store information contained in measurements of system state. Proof that this is the case is readily apparent from the good fit that can often be obtained between these measurements and corresponding model outputs as the model is calibrated. The problem, however, is that the receptacles, which hold this information are somewhat corrupted from an expert knowledge point of view. As will now be discussed, this may have important adverse consequences for some model predictions.

7.2.6 Consequences of model simplicity

All models are simplifications of reality, even those with 'realistic' parameter fields that strive to reflect geological expert knowledge. All models employ a grid, a mesh, or analytical equations in place of complex system geometries and processes that operate on multiple scales in highly heterogeneous media. Furthermore, no parameter field, no matter how much it attempts to reflect the spatial nuances of real-world geological heterogeneity, bears more than a passing resemblance to reality. In any model, stresses and boundary conditions and their variability in space and time, are represented only approximately. This applies especially to recharge processes and to the interaction of a groundwater system with surface water systems. If we continue to view a model and its parameters as receptacles for information, pertaining to a complex natural system, it follows that when the parameters of a defective model are adjusted in order to fit its outputs to that system's behaviour. The information contained in that behaviour is thereby directed to receptacles that are corrupted by the model's defects.

How much does this matter? This is a difficult question to answer. It is also a question whose answer is prediction-specific.

Using theory inspired by subspace analysis, Doherty & Welter (2010), Doherty & Christensen (2011), Watson et al. (2013), and White et al. (2014) have inquired into the effects of model simplicity on the history-matching and prediction processes. They define 'optimal simplification' of a model as that, which incurs the least cost to a model's parameters as they are adjusted to respect history-matching constraints. 'Cost' is equated to the propensity for the history-matching process to induce bias in estimated parameters of a defective model. This occurs when the values estimated for at least some model parameters are 'off-centre' with respect to their true posterior probability distributions, sometimes so much off-centre as to deviate considerably from their true (but unknown) values. Paradoxically, parameter bias thus incurred may or may not be transferred to predictions made by the model. Hence for some predictions, model imperfections can be 'calibrated out'. However, for other predictions the situation is very different. The above authors show that in certain circumstances it is possible for some model predictions to exhibit little or no bias prior to history-matching, but to exhibit substantial posthistory-matching bias. This can occur even though a model may be complex enough and be endowed with enough parameters for the history-matching process to yield a good fit between model outcomes and historical measurements of system state. Sadly, the amount of bias that may be introduced through history-matching cannot be quantified.

The studies also show that where structural defects of a model are exposed via the history-matching process through a demonstrable incapacity of the model to compute outcomes that fit all historical data, it may still be possible to fit some aspects of that data in a way that does not induce parameter and predictive bias. They show that through astute formulation of the calibration objective function (normally a weighted least squares measure of misfit between raw and processed field data on the one hand and corresponding raw and processed model outputs on the other hand), the information content of a measurement dataset may be prevented from entering parameter receptacles that are excessively corrupted by model simplicity. In some modelling contexts the same set of field data may need to be subjected to more than one type of processing before being matched to model outcomes that are processed in the same ways. This can occur where the same data contains information that informs more than one aspect of the system simulated by the model. For example, temporal head differences are rich in information on recharge processes and aquifer storage capacity, while vertical head differences are informative of aquitard flow resistance. A model can often compute these head differences with less structural corruption than absolute heads. It is these model-calculated differences that should then be matched to corresponding differences in measured heads at the same time as unprocessed model-generated and measured heads are matched. An appropriate weighting strategy must then be adopted to ensure visibility of these head differences in a final, multi-component objective function that reflects these and other types of processed and raw measurement data.

Unfortunately, however, the outcomes of model simplification are often invisible to the history-matching process. This is especially the case where calibration is undertaken through highly-parameterised inversion, wherein dense spatial parameterisation, supported by appropriate mathematical regularisation, is employed to allow the calibration process to respond to information on hydrogeological heterogeneity, contained in the head and other data that collectively comprise a calibration dataset. A good fit between measurements and corresponding model outputs may thereby be obtained. Estimated parameter values may be reasonable. However, the above studies show that this does not protect parameters from history-matching-induced bias where other aspects of the model are over simplified or incorrect, for example boundary conditions and stresses. Some decision-critical model predictions may then inherit this bias while others may not. As stated above, where predictive bias is thus incurred, its magnitude cannot be quantified, compensated for, nor included in model predictive uncertainty assessment as 'man-made uncertainty'. As for its visible, simplification-induced, 'structural noise' counterpart, some protection against its effects may be afforded through formulation of a strategic, multi-component objective function. Alternatively (or as well), its deleterious effects on some predictions can also be mitigated by deliberately seeking a mediocre fit between model outputs and field measurements. However, given that other predictions are inoculated against history-matching-induced predictive bias by virtue of the nature of their sensitivities to model parameters, this strategy must be used with caution and should be adopted on a prediction-specific basis.

All of this has important consequences for the way that models are constructed, calibrated, and deployed in the decision-making context. Some of these consequences will be addressed in more detail below. Meanwhile, the above discussion can be summarised as follows. Where a model is a simplified version of reality (as all models are), the receptacles that it and its parameters offer for the various types of information that are resident within historical measurements of system state become corrupted. The extent of this corruption may not be such as to prevent that model from replicating all components of a measurement dataset when subjected to parameter adjustment through the history-matching process. The model may thus assume the mantle of being 'well-calibrated'. However, the cost of being 'well-calibrated' may be a substantial, though unquantifiable, propensity for bias in some of its predictions. At the same time, however, the history-matching process may have engendered no bias in other model predictions, while reducing the uncertainties of those predictions dramatically. These latter predictions tend to be those that bare most resemblance to the measurements against which the model is calibrated and are made when the model operates under a similar stress regime to that under which it operated historically. In contrast, those predictions that are at most risk of sustaining bias are those that are sensitive to parameters that are partially informed by the measurement dataset, but must also be partially informed by expert knowledge because of an information deficit with respect to these parameters in the measurement dataset. These are the parameters that, if called upon to do so, can most readily assume roles that compensate for model inadequacies, as they are adjusted in order for model-computed outputs to match measurements that comprise a calibration dataset.

Because predictive bias cannot be quantified in most contexts of model deployment, the efficacy of different bias mitigation strategies can only be qualitatively judged. Nevertheless experience, intuition, and the outcomes of research, such as that discussed above can provide some guidance. The model development and history-matching process thus becomes qualitative, subjective, and prediction-specific.

7.2.7   Models and modellers

So where does the above discussion leave 'the calibrated model' as the cornerstone of groundwater management? What are the repercussions for the more advanced notion of a scientific instrument that is used to test and maybe reject hypotheses that bad things will happen?

An immediate outcome is that a groundwater model (or any other environmental model for that matter) whose primary role is that of decision-support must be purpose-built. This purpose is the testing of a specific hypothesis concerning the happening of a specific unwanted outcome of a specific management strategy. The nuances of its construction, with the gross simplification of reality that this entails, must be tuned to that purpose. The decision to embrace expert knowledge on the one hand or a calibration dataset on the other hand as the principle agent for hypothesis rejection must also be prediction-specific. Where hypothesis rejection is attempted through demonstrating incompatibility of the tested hypothesis with historical system behaviour, the grounds for this rejection are qualitative and once again prediction-specific.

The notion then that a government or private agency can commission the building and calibration of a single model whose domain encompasses a large area, and that this model can then be delivered to another party in order to make a broad range of predictions, pertaining to a broad range of environmental management options, pertaining to that area is highly questionable. Nevertheless, this mode of model construction and deployment is commonplace. In promulgating the perpetuation of this mode of model usage, those who pay for such models and stakeholder groups whose interests are affected by decisions based on such models, are convinced of the integrity of the modelling process through recourse to the fact that a model was constructed in accordance with a set of guidelines, or standards. Rarely, if ever, however, do the guidelines, which are adopted by our industry draw attention to the fact that a model's place in the decision-making process should be one of testing whether a specific unwanted outcome of a favoured management practice is demonstrably incompatible with expert knowledge of a system and of the historical behaviour of that system. Nor do they note that this hypothesis testing can only be qualitative and that it is best performed with a purpose-built model, constituting a scientific instrument which is tuned to this particular task.

It should be apparent from the above discussion that the model construction and history-matching processes are fraught with many compromises. In the interests of good environmental management it is obvious that these compromises must be made by personnel who are equipped with the knowledge necessary to make them and have no bias towards a particular outcome. This knowledge must include not just that of processes and properties that prevail within a study area and of the numerical tools that seek to simulate those processes. It must also include an understanding of the concepts and operational details of geostatistical, data-processing, and inversion software packages with which a model must interact if it is to become a repository of the information that is necessary for the testing of hypotheses on which environmental decision-making depends.

This blending of advanced numerical processing with informed subjectivity requires that modelling be considered as a component of an overall decision-making process, undertaken by a team of experts of which the modeller is a member. Modelling should not be considered as an isolated activity that can be 'contracted out' and whose 'deliverable' is the production of a tool that can then be used independently of those who built it in order to predict the future behaviour of a system that may be subjected to stresses that are very different from those which it has experienced in the past. With the modeller viewed as part of the decision-making team, the outcomes of the modelling process are likely to be more than one model, each tuned to illuminate different aspects of what may be a complex set of factors to consider before a final decision can be made. None of these models should claim to be a 'good simulator' of a complex environmental system, for such a claim is too audacious for any model. However, each model may claim special status as an optimal repository of the knowledge and information that is necessary for the testing of one specific hypothesis that, along with other hypotheses, must be tested as part of an overarching decision-making process.

To the skills required of modellers must be added that of convincing those who pay for models that this is what they should be paying for. The impetus for change will not come from those who do not understand what a model can and cannot achieve. It must come from modellers themselves who as scientists must insist that the tools which they develop are used for the sole purpose of supporting an unbiased decision-making process that gives full voice to the scientific method. The integrity of our profession, and intergenerational equity in natural resource management demand this.

7.3   DISCUSSION AND CONCLUSIONS

This chapter has attempted to present a chain of logic leading to the definition of a suitable place for groundwater modelling in groundwater management. It has attempted to recast models less as near perfect simulators of environmental behaviour, than as repositories of environmental information—information born of expert knowledge and information residing in measurements of system behaviour. Its role in the decision-making process is that of employing this information to assess the likelihood of occurrence of one or a number of unwanted outcomes of groundwater system management. It does this through testing whether the occurrence of these events is compatible with this information. In performing this task it can provide indispensable support for the development of a management strategy that can guarantee, at a high level of confidence, that bad things, which people do not want to happen, will not in fact occur. This mode of model usage satisfies the requirements of the decision-making process at the same time as it implements the scientific method and maintains scientific integrity of modellers. It aligns the expectations of modelling with the nature of expert knowledge together with its interplay with measurements of system state as expressed by Bayes equation. It acknowledges that a model can claim no ability to predict what will happen in the future if a certain management strategy is adopted. However, it can claim an ability to predict what will not happen in the future, following adoption of that strategy.

Failure for a modelling exercise is the incursion of a type II statistical error. This occurs if a hypothesis is falsely rejected. In the decision-making context it occurs if something bad happens, following a promise that it will not happen. Though awareness of this error may not arise for years, the integrity of a modelling exercise should be judged on the basis of whether a modeller has provided the necessary assurances that this type of error has been avoided. Standards and guidelines whose intention is to assist modellers to engage in the modelling process and decision makers to understand the modelling process should address this issue above all others.

If a model is to take its logical place in the decision-making process based on the above precepts, full account must be taken of its imperfections. Of particular importance is the fact that, with present modelling technology, the two terms on the right side of Bayes equation compete with each other at the same time as they complement each other in quantifying and reducing the uncertainties associated with predictions of management interest. With present technology, enhancement of a model's ability to encapsulate expert knowledge can erode its ability to encapsulate information that resides in measurements of system state and vice versa. A modeller must therefore decide on the type of information which they would like his/her model to best encapsulate. The cost of incomplete encapsulation of the other type of information must then be taken into account through inflated estimates of predictive uncertainty. This may then affect the confidence level with which decision-critical hypotheses are rejected, if incursion of a type II statistical error is to be avoided.

Caution, expertise, and integrity are required when attempting to introduce the information to a model that is resident in measurements of system state through the history-matching process. Field data and corresponding model outputs may need to be processed in strategic ways before being compared with each other through a multi-component objective function and then matched with each other through reduction of that objective function. A qualitative weighting strategy that ensures that each objective function component is not dominated by other objective function components in the overall objective function may need to be adopted. Though subjective, such a strategy is required for the reduction of simplification-induced 'structural noise' and for minimisation of history-matching-induced predictive bias.

At the same time it must be borne in mind that the effect of history-matching on predictive bias is prediction-specific. For some predictions 'any calibration is good calibration'. This is because these predictions are sensitive to model parameters in the same way that members of the measurement dataset are sensitive to model parameters. The information content of a measurement dataset is then directly transferred to these predictions in spite of the defective nature of parameters as receptacles of expert knowledge though which this measurement information must pass. In order to facilitate this transfer of information, parameters may need to adopt values that expert knowledge would deem to be lacking in credibility as a high level of fit between model outputs and field observations is achieved. For other predictions the opposite is the case, as even mild history-matching can induce considerable predictive bias. In this case the history-matching process may do more harm than good if more than a mediocre fit between model outputs and corresponding field observations is sought. Consequently, mild constraints on parameter values should be imposed through the history-matching process if a type II statistical error is to be avoided.

Where a decision-significant prediction is very different in character from observations, comprising the history-matching dataset and where model-based uncertainty analysis suggests that the uncertainty of that prediction is lowered by only a small amount through the history-matching process, it is possible that the potential for predictive bias incurred by the history-matching process may be greater than the reduction in uncertainty that could potentially be accrued through it. Sadly, this potential for post-history-matching bias cannot be assessed by the model itself. The possibility of its presence and the magnitude of its effect can only be inferred from the circumstances associated with a particular modelling context. In a case such as this, model predictive uncertainty analysis should be based on expert knowledge alone. This may be achieved through engaging the model in standard Monte Carlo analysis, in which geostatistical realisations of model parameter fields are generated and a prediction of interest is computed based on each one of these. It is noteworthy that few, if any, of these realisations are pertinent model outputs likely to match historical measurements of system state (though the modeller should assure themselves that these outputs collectively span the measurement dataset). If the purpose of the modelling exercise is the analysis of the uncertainty associated with a decision-critical prediction (as it should be), then this should not be construed as invalidating that analysis, as it is the predictive probability distribution that is important and not the values of individual predictions that sample that distribution. In contexts where the measurement dataset is information-poor with respect to a prediction of management interest, Bayes equation demonstrates that, had it been possible to calculate the posterior probability distribution of that prediction instead of its prior predictive probability distribution, then the latter would have been narrower than the former. A type II statistical error is thereby forestalled.

Groundwater modelling has come a long way in the last 30 years. It has been welcomed by environmental managers into the inner sanctum of the decision-making process. However, the role that it should play in that process needs to be refined if it is to provide what managers actually need (as distinct from what they think they need), while remaining true to its primary task of implementing the scientific method. Modelling software is getting better all the time. This is a good thing. However, more than this is required if modelling is to realise its full potential in decision support.

Realisation of this potential can only be achieved through recognition of the fact that a model can achieve little on its own. Instead, it must be recognised that a model needs a partner—perhaps a number of partners. These must include geostatistical, inversion, and uncertainty analysis software packages that are of equal sophistication to that of the model. Without these packages the passage of information into and out of the model falls vastly short of the capacity of the model to hold such information. The impetus for further development of these packages must be based on an insistence by the modelling community that they are essential for it to do its job properly. At the same time, the acquisition of the skill set and knowledge base that is required for the development and use of these packages must be a matter of highest priority for the industry, including the educational and research institutions that support the industry.

REFERENCES

Aster R.C., Borchers B., Thurber H. (2005) *Parameter estimation and inverse problems*. 301 pp. Elsevier, Amsterdam, Netherlands.

Beven, K. (2009) *Environmental modelling: An uncertain future?* Routledge, London.

Certes C., de Marsily G. (199) Application of the pilot points method to the identification of aquifer transmissivities. *Advances in Water Resources* 15(5), 284–300.

Costanza R., Kubiszewski I. (2012) The authorship structure of "ecosystem services" as a transdisciplinary field of scholarship. *Ecosystem Services* 1, 16–25.

Deutsch C.V., Journel A.G. (1998) *Geostatistical software library and user's guide*. Second Edition, Oxford University Press, Oxford.

Doherty J. (2003) Groundwater model calibration using Pilot Points and Regularisation. *Ground Water* 41(1), 170–177.

Doherty J. (2014) PEST: Model-independent parameter estimation. Watermark Numerical Computing, Brisbane.

Doherty J., Christensen S. (2011) Use of paired simple and complex models to reduce predictive bias and quantify uncertainty. *Water Resources Research* 47(12), W12534.

Doherty J., Welter D. (2010) A short exploration of structural noise. *Water Resourources Research* 46, W05525, doi:10.1029/2009WR008377.

Freeze R.A., Massmann J., Smith L., Sperling T., James B. (1990) Hydrogeological decision analysis: 1 A framework. *Ground Water* 28(5), 738–766.

Keating E., Doherty J., Vrugt J.A., Kang Q. (2010) Optimization and uncertainty assessment of strongly nonlinear groundwater models with high parameterization dimensionality. *Water Resources Research* 46, W10517, doi:10.1029/2009WR008584.

Laloy E., Vrugt J.A. (2012) High-dimensional posterior exploration of hydrologic models using multiple-try DREAM(Z,S) and high-performance computing. *Water Resourources Research* 48(1), doi:10.1029/2011WR010608.

Moore C., Doherty J. (2005) The role of the calibration process in reducing model predictive error. *Water Resources Research* 41(5), W05020, doi:10.1029/2004WR003501.

Moore C., Doherty J. (2006) The cost of uniqueness in groundwater model calibration. *Advances in Water Resources*. 29(4), 605–623.

Moore C., Wöhling T., Doherty J. (2010) Efficient regularization and uncertainty analysis using a global optimization methodology. *Water Resourources Research* 46, W08527, doi:10.1029/2009WR008627.

Remy N., Boucher A., Wu J. (2009) *Applied Geostatistics with SGEMS. A User's Guide.* Cambridge University Press, Cambridge.

Tonkin M., Doherty J. (2005) A hybrid regularised inversion methodology for highly parameterised models. *Water Resources Research* 41, W10412, doi:10.1029/2005WR003995, 2005.

Tonkin M., Doherty J. (2009) Calibration-constrained Monte Carlo analysis of highly parameterized models using subspace techniques. *Water Resources Research* 45, W00B10, doi:10.1029/2007WR006678.

Wallace K.J. (2007) Classification of ecosystem services: Problems and solutions. *Biological Conservation* 139(3–4), 235–246 http://dx.doi.org/10.1016/j.biocon.2007.07.015.

Watson T.A., Doherty J.E., Christensen S. (2013) Parameter and predictive outcomes of model simplification. *Water Resources Research* 49 (7), 3952–3977.

Welter D.E., Doherty J.E., Hunt R.J., Muffels C.T., Tonkin M.J., Schreüder, W.A. (2012) Approaches in highly parameterized inversion: PEST++, a Parameter ESTimation code optimized for large environmental models. U.S. Geological Survey Techniques and Methods, book 7, section C5, 47 p.

White J.T., Doherty J.E., Hughes J.D. (2014) Quantifying the predictive consequences of model error with linear subspace analysis. *Water Resources Research* 50(2), 1152–1173.

Zhou H., Gomez-Hernandezm J.J., Li L. (2014) Inverse methods in hydrogeology: evolution and recent trends. *Advances in Water Resources* 63, 22–37.

Subject index

Series IAH-Selected Papers

Printed and bound by CPI Group (UK) Ltd, Croydon, CR0 4YY

24/10/2024

01778286-0011